IFSTA 105
SIXTH EDITION

FIRE STREAM PRACTICES

VALIDATED BY

THE INTERNATIONAL FIRE SERVICE TRAINING ASSOCIATION

PUBLISHED BY

FIRE PROTECTION PUBLICATIONS • OKLAHOMA STATE UNIVERSITY

EDITORIAL STAFF
Gene P. Carlson - Editor
Charles W. Orton - Assistant Editor
Don Davis - Production Coordinator
Cover photo by E. Ann Carlson

ifsta

Dedication

This manual is dedicated to the members of that unselfish organization of men and women who hold devotion to duty above personal risk, who count sincerity of service above personal comfort and convenience, who strive unceasingly to find better ways of protecting the lives, homes and property of their fellow citizens from the ravages of fire and other disasters . . . The firefighters of All Nations.

Dear Firefighter:

The International Fire Service Training Association (IFSTA) is a nonprofit organization that exists for the sole purpose of serving firefighters. IFSTA is a member of the Joint Council of National Fire Organizations, National Fire Protection Association and International Society of Fire Service Instructors. If you need help in locating additional information concerning the organization, training materials, or manual orders, please contact one of the people listed below.

Without you we do not exist!

Write:
Customer Services
Fire Protection Publications
IFSTA Headquarters
Oklahoma State University
Stillwater, Oklahoma 74078

or call:
(405) 624-5723
Manual Orders — Brenda Dale

ISBN 0-87939-041-7
Library of Congress 80-80447
Sixth Edition
Printed in the United States of America

THE INTERNATIONAL FIRE SERVICE TRAINING ASSOCIATION

The International Fire Service Training Association is an educational alliance organized to develop training material for the fire service. The annual meeting of its membership consists of a workshop conference which has several objectives —

. . . to develop training material for publication
. . . to validate training material for publication
. . . to check proposed rough drafts for errors
. . . to add new techniques and developments
. . . to delete obsolete and outmoded methods
. . . to upgrade the fire service through training

This training association was formed in November 1934, when the Western Actuarial Bureau sponsored a conference in Kansas City, Missouri, to determine how all agencies that were interested in publishing fire service training material could coordinate their efforts. Four states were represented at this conference and it was decided that, since the representatives from Oklahoma had done some pioneering in fire training manual development, other interested states should join forces with them. This merger made it possible to develop nationally recognized training material which was broader in scope than material published by an individual state agency. This merger further made possible a reduction in publication costs, since it enabled each state to benefit from the economy of relatively large printing orders. These savings would not be possible if each individual state developed and published its own training material.

From the original four states, the adoption list has grown to forty-four American States; six Canadian Provinces; the British Territory of Bermuda; the Australian State of Queensland; the International Civil Aviation Organization Training Centre in Beirut, Lebanon; the Department of National Defence of Canada; the Department of the Army of the United States; the Department of the Navy of the United States; the United States Air Force; the United States Bureau of Indian Affairs; the United States General Services Administration, and the National Aeronautics and Space Administration (NASA). Representatives from the various adopting agencies serve as a voluntary group of individuals who govern policies, recommend procedures, and validate material before it is published. Most of the representatives are members of other international fire protection organizations and this meeting brings together men from several related and allied fields, such as:

. . . key fire department executives and drillmasters,
. . . educators from colleges and universities,
. . . representatives from governmental agencies,
. . . delegates of firefighter associations and organizations, and
. . . engineers from the fire insurance industry.

This unique feature provides a close relationship between the International Fire Service Training Association and other fire protection agencies, which helps to correlate the efforts of all concerned.

The publications of the International Fire Service Training Association are compatible with the National Fire Protection Association's Standard 1001, "Fire Fighter Professional Qualifications (1974)," and the International Association of Fire Fighters/International Association of Fire Chiefs "National Apprenticeship and Training Standards for the Fire Fighter." The standards are an effort to attain professional status through progressive training. The NFPA and IAFF/IAFC Standards were prepared in cooperation with the Joint Council of National Fire Service Organizations of which IFSTA is a member.

In order to formally adopt IFSTA training material, the firefighters should first request it through their training agencies. While adopting does not obligate any sponsoring agency, it still profits by sending a representative to the Annual IFSTA Validation Conference. This representation permits a voice and a vote on the ever-changing policies, techniques, and procedures which must be reflected in each publication. Adoption further adds prestige to any training program and this, in turn, adds prestige to IFSTA publications.

The International Fire Service Training Association meets each July at Oklahoma State University, Stillwater, Oklahoma. Fire Protection Publications at Oklahoma State University publishes all IFSTA training manuals and texts. This department is responsible to the executive board of the association. While most of the IFSTA training manuals can be used for self-instruction, they are best suited to group work under a qualified instructor.

PREFACE

Although one would think that with something as basic as hydraulics there could be nothing new, this sixth edition of IFSTA 105, *Fire Stream Practices* proves otherwise. At least insofar as the fire service is concerned.

New methods and materials used in manufacturing fire hoses have made the $2Q^2 + Q$ friction-loss formula inapplicable in some cases. The flow meter has finally found a place on the pump panel, freeing the pump operator from having to make friction-loss calculations. The growing acceptance of the metric system of measurement is pushing the fire service in the United States into thinking more seriously about how this will affect fire suppression equipment and operations.

In an effort to reflect these changes, this edition of *Fire Stream Practices* includes much new material and has been extensively revised. For example, metric equivalents have been inserted where appropriate. Where parenthetical equivalents would be inappropriate for illustrations, new illustrations and explanations have been devised for the assistance of our readers who already use the metric system.

Material has been updated or newly incorporated where necessary to help the firefighter meet the requirements of NFPA Standard No. 1001 and 1002. Also, to aid the firefighter in gaining proficiency in making hydraulics calculations, Chapter 3 has calculation exercises.

To assist the reader in finding information quickly, each chapter divider has a bleed screen at the edge, and an index has been added.

This sixth edition of *Fire Stream Practices* could not have been so extensively revised without the assistance of many people. First, and most, credit must go to the members of the validation committee:

Chairman
Arlen Gross
Louisiana State University
Baton Rouge, Louisiana

Vice Chairman
W. S. "Stan" Thompson
Elkhart Brass Manufacturing Co., Inc.
Elkhart, Indiana

Secretary
Nick Renihan
Akron Brass Co.
Wooster, Ohio

Len Gatin
City of Kamloops Fire Dept.
Kamloops, British Columbia

Don Seelig
Oklahoma State University
Stillwater, Oklahoma

C. H. McMillan
Fire Task Force Innovations, Inc.
Valparaiso, Indiana

Jack Whisman
Longmont Fire Dept.
Longmont, Colorado

We also express our thanks to the other loyal and hard-working committee members whose talents and knowledge made this manual possible:

Elmo Anderson	Dale Scobee
Charlotte Badgett	Robert Bloom
Raymond E. Linville	Eric Haussermann
Ray Gish	Joseph Donovan
Dave Ballenger	Don Froese
Mike Gray	Elsworth Greer
Robert Ely	Edward Sniechoski

The following generously granted permission to reprint copyrighted material or provided photographs and illustrations:

Akron Brass Co.
American Insurance Association
Elkhart Brass Manufacturing Co., Inc.
Emergency One, Inc.
Warren E. Isman
National Fire Protection Association
Task Force Tips
Waterous Company
Insurance Services Offices
Electrical World

Our gratitude is also extended to the following individuals whose fine work gave this manual its finished form: Ann Moffat, artwork, layout, pasteup; Nancy Wilson, phototypesetting; Pat Knight, research; and Stan Hazen, proofreading.

CONTENTS

INTRODUCTION

HYDRAULICS AND FIRE STREAMS

Hydraulics in its broadest sense is an exact science that deals with water and other fluids while at rest and in motion. A liquid does not need to be encased within a tube, pipe or hose to comply with this definition, since the flow of water in rivers and canals is also a hydraulic study. A fire stream is a stream of water shaped by nozzle design. Its application to fire extinguishment has established the term *fire stream*.

Fire Stream Practices is a study concerning the actions required of firefighters when they effectively manipulate hose, nozzles and appliances to apply water in a form best suited to extinguish fire. Water has always been the most practical and commonly used agent for extinguishing fire. With the development of fire hose for transporting water and pumps for developing pressure, water could be forced through a hose line with sufficient speed to carry it from a nozzle through space to a desired point. Before 1888 there were no exact standards or data that controlled the construction of nozzles nor were there any satisfactory or dependable tables relating to fire streams. In that year a series of experiments with pressures ranging from 5 to 100 pounds per square inch (35 to 700 kPa) at the nozzle were made by John R. Freeman of the American Society of Civil Engineers. Freeman's experiments used several kinds, shapes and sizes of nozzles. This comparison of nozzles was an endeavor to find a design that would produce the best water jet. The Freeman experiments were the most accurate and complete ever undertaken on the subject and the fire stream tables developed by him are still used as a standard.*

1888 — Freeman's tests basis for today's calculations

---

*Frequently, in attempting to duplicate data contained in this manual, individuals may presume the data in this manual to be inaccurate because of different data recorded by their tests. Unless tests are performed under conditions identical to those from which the data in this manual was developed, there will be differences. These differences may be attributed to many factors such as the development of more modern equipment and materials for hose, fittings and nozzles; improper use or lack of adequately calibrated measuring equipment; and use of different hose materials that may allow for greater expansion, increasing the inside diameter. Where extreme accuracy is necessary, tests may have to be performed to develop data similar to that contained in this manual; however, the data contained in this manual is considered correct for use by the fire service unless proven otherwise.

During the early Freeman experiments and tests the major objective was to determine the best form of nozzle and playpipe for practical purposes. "Practical purposes" at that time meant to find out what form of nozzle would at a given pressure project a given number of gallons of water per minute to the greatest height or distance. This feature required a compact stream with the least loss from detaching drops and spray.

Today's nozzles cause less water damage

Since that time the design of fire stream nozzles has been improved to more effectively extinguish fire with a minimum amount of water damage. Manufacturers have designed fire appliances that function with a reduced amount of resistance to the flow of water and nozzles that permit a selection of the type of stream desired. The transition to the use of devices to control fire by the widespread use of water in the form of droplets resulted in the development of effective spray and fog nozzles. These nozzles have replaced solid stream nozzles to a great degree, but solid streams are still needed to perform definite functions in the fire service.

An extensive study of the application of water fog on fires within confined spaces was conducted by an exploratory committee representing the Fire Department Instructors Conference and the International Fire Service Training Association. The purpose of this study was to devise effective and practical methods of controlling and extinguishing fires. Full-scale fire tests on buildings were conducted to determine the effect of the indirect method of fire extinguishment and to encourage a study of the need for further development of fog nozzle design. Modern fire fighting methods involve the application of water in a form that will quickly reduce the temperature of burning material to a point where combustion no longer takes place. Field experience and study by many fire officials have conclusively demonstrated that the application of water on fires in other than a solid stream possesses great merit for the control of certain types of fires.

DEFINITION, PURPOSE AND SCOPE

From the foregoing discussion, a fire stream can be defined to be a stream of water after it leaves the fire hose and nozzle until it reaches the desired point. The perfect fire stream can no longer be sharply defined, since individual desires and extinguishing requirements vary. During the time a stream of water passes through space it is influenced by its velocity and by gravity, wind and friction with the air. The condition of the stream when it leaves the nozzle is influenced by operating pressures, nozzle design, nozzle adjustment and the condition of the nozzle orifice.

Fire stream's purpose: reduce temperature

The purpose of this manual is the explanation of the purposes of fire streams and their production. Fire streams are intended to reduce temperatures and provide protection by one of the following methods:

- Applying water directly to the burning material.
- Reducing high atmospheric temperature and absorbing and/or dispersing hot smoke and fire gases from a heated area in a confined space.
- Reducing the temperature over an open fire, thus permitting a closer approach with hand hoselines to effect extinguishment.
- Being used as a heat-reflecting and -absorbing shield.

The study and knowledge of the material in this manual is not only important to the firefighter at the pump operator level, it is also helpful to chief officers or others in need of information to aid them when making decisions pertaining to the selection and purchase of fire fighting equipment.

The scope of this manual is an attempt to fulfill the foregoing purposes by establishing the requirements, principles and characteristics of fire streams; by providing methods to produce or develop fire streams; and by presenting a study of related information for the selection and application of fire streams.

1

FIRE AND WATER

NFPA STANDARD 1001

Firefighter I

3–8 Fire Streams.

3–8.1 The firefighter shall define a fire stream.

3–8.2 The firefighter, given the necessary resources, shall manipulate the nozzle so as to attack at least two live fires including: a Class A fire and a Class B fire.

3–8.3 The firefighter shall define water hammer and at least one method for its prevention.

3–8.4 The firefighter shall demonstrate how to properly open and close a nozzle.

Firefighter II

4–8 Fire Streams.

4–8.1(a) The firefighter, given fire situations, for each situation shall identify the phase of burning.

NFPA STANDARD 1002

3–4 Hydraulic Calculations.

3–4.1 The fire apparatus driver/operator shall demonstrate the principles of friction loss as they relate to:

(a) Internal diameter of hose

(b) Length of hose line

(c) Manner in which hose lines are laid

(d) Physical condition of hose

(e) Pressure . . .

(i) Velocity of flow

3–4.2 The fire apparatus driver/operator shall identify the following types of fluid pressure encountered in the fire service:

(a) Flow pressure

(b) Negative pressure

(c) Normal operating pressure

(d) Residual pressure

(e) Static pressure

3–4.3 The fire apparatus driver/operator shall identify the following terms that relate to the basic principles of fire service hydraulics:

(a) Atmospheric pressure . . .

(e) Friction loss

(f) Head pressure (back pressure) . . .

(j) Pounds per square inch . . .

(l) Vacuum

(m) Velocity

(n) Water hammer

FIRE AND WATER

Firefighters most frequently have to extinguish Class A fires, those involving ordinary combustibles (wood, paper, textiles, etc.). The most frequently used fire extinguishing agent for these fires is water. Water is not only the most plentiful of all fire extinguishing agents, but also the most easily handled and, because of its nature, the best extinguishing agent for these fires. The reasons for this can be understood most clearly with a review of fire chemistry and behavior, along with a study of the natural characteristics of water.

STRUCTURAL FIRES — PHASES OF BURNING

Structural fires are of two types: open and confined. The open fire is one that has broken through to burn freely in the outside air. Confined fires are those in which the heat, smoke and gases are still largely confined in the interior of a structure. Confined fires develop in three stages known as *the three phases of fire:* incipient, free-burning and smoldering.

In the first, or incipient phase, oxygen is available and the fire is producing water vapor (H_20), carbon dioxide (CO_2), and perhaps a small quantity of sulfur dioxide (SO_2). There will be some carbon monoxide (CO), but it is negligible at this stage. Some heat is being generated and the amount increases with the progress of the fire. The fire may be producing a flame temperature of around 1,000°F (538°C). At this point the oxygen content of the room is approximately 20 percent.

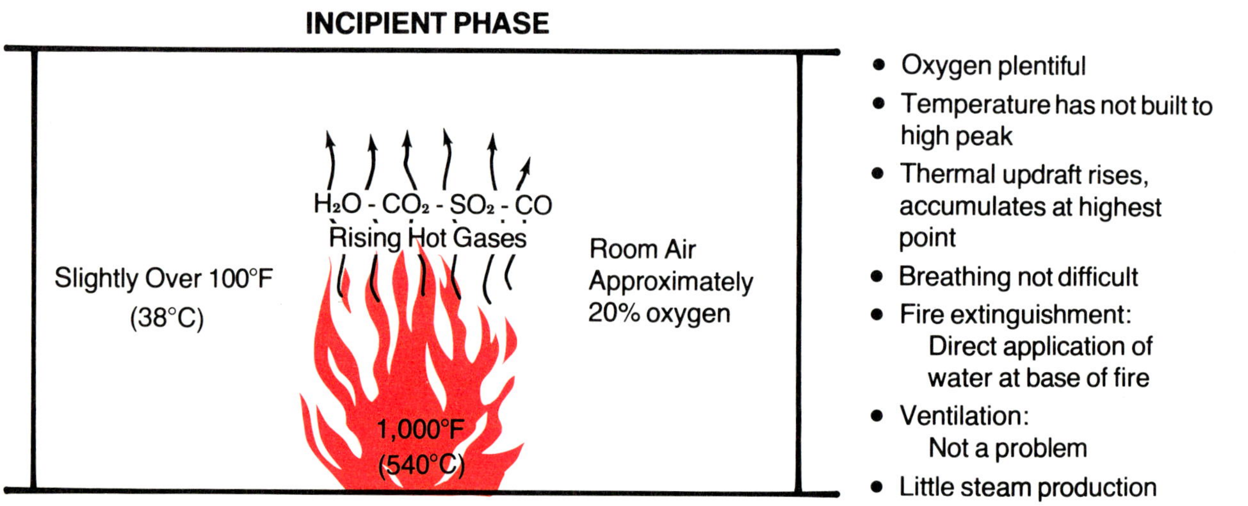

Figure 1.1 — The incipient, or beginning, phase of a fire.

In the second, or free-burning phase, the oxygen in the room has been reduced, but enough remains to support a free-burning fire. All of the combustible material in the upper levels of the room will have been heated to their ignition temperatures because of the rapid generation of heat, and they may be burning violently. The transfer of heat by convection makes the upper levels the hottest. The heated condition of the atmosphere causes rapid expansion of the gases. This second phase is a period of rapid burning, and entrance may be impossible because of the intense heat and rapidly increasing amount of smoke and fire gases.

FREE BURNING PHASE

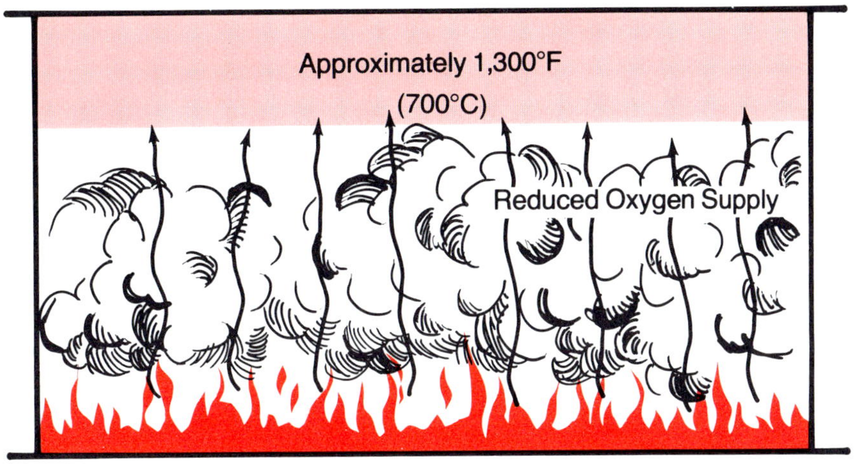

- Fire has involved more fuel
- Oxygen supply is being depleted
- Heat accumulates at upper areas
- Breathing difficult:
  Masks recommended
- Fire extinguishment is reaching the area of major involvement
- Ventilation:
  Not a definite need
- Good steam production

Figure 1.2 — The free-burning phase of a fire. Oxygen supply is decreased and temperature increases in the room.

In the third, or smoldering phase, the fire is barely visible because the reduction of oxygen has increased the amount of carbon monoxide, free carbon and unburned fuel gases. The fire will smolder and the room will completely fill with dense smoke and gases of combustion at a temperature of well over 1,000°F (538°C).

SMOLDERING PHASE

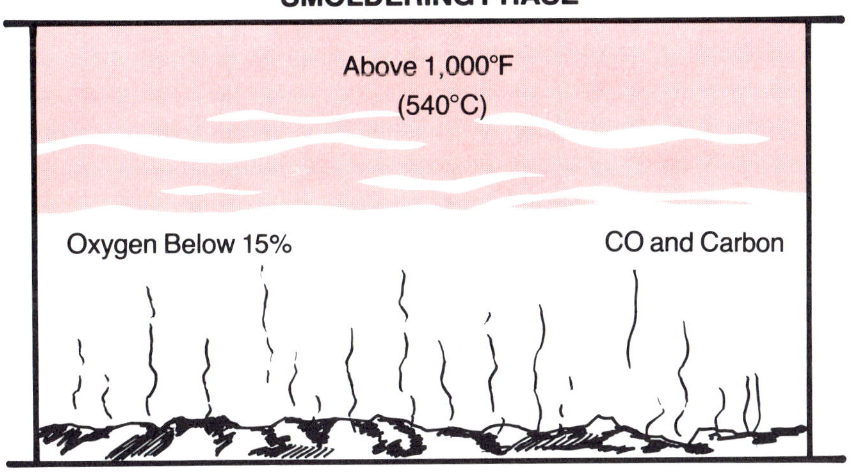

- Oxygen supply not equal to demands of fire
- Temperature throughout building is very high
- Normal breathing is not possible
- Oxygen deficiency may cause backdraft
- Fire extinguishment
  Indirect method
- Ventilation:
  A must!
- Maximum steam production from water fog

Figure 1.3 — The smoldering phase. Oxygen content below 15%; flame may cease, but dense smoke and heat completely fill room.

THE FIRE TETRAHEDRON

For a fire to exist there must be a combination of four factors: fuel, heat, oxygen and a chemical chain reaction. Taking any of these four components from a fire will extinguish the fire. This basic observation gives the four methods of fire extinguishment:

- Remove the fuel (separation)
- Remove the heat (cool)
- Remove the oxygen (smother)
- Prevent the chain reaction (interfere with chemical action)

The most easily accomplished of these four methods is cooling with water: separating the fuel from the flames of a structural fire is nearly impossible; smothering is, also; and the cost of chemicals to prevent the chain reaction would be prohibitive.

THE CHARACTERISTICS OF WATER

Because water is the most frequently used extinguishing agent the firefighter must know its characteristics and the principles involved to transport it through pipe, fire hydrants, pumps, hose and nozzles.

Water is a compound of hydrogen and oxygen formed when two parts of hydrogen combine with one part of oxygen; hence, the formula for water is H_20. When hydrogen burns in the presence of oxygen (air) a chemical reaction takes place that produces water vapor. This water vapor (gas) will condense into liquid when it cools. Water at normal temperatures (32° to 212°F, 0° to 100°C) exists in abundant quantities. Below 32°F (the freezing point of water) it converts to the solid state in the form of ice. Above 212°F (the boiling point of water) it converts into a gas called *water vapor* or *steam*. In this vapor form, water cannot be seen. When steam starts to cool, however, its visible form is called *condensed steam*. Figure 1.4 illustrates the physical state of water at its three transitional temperatures.

Figure 1.4 — The three physical states of water, with transformational temperatures.

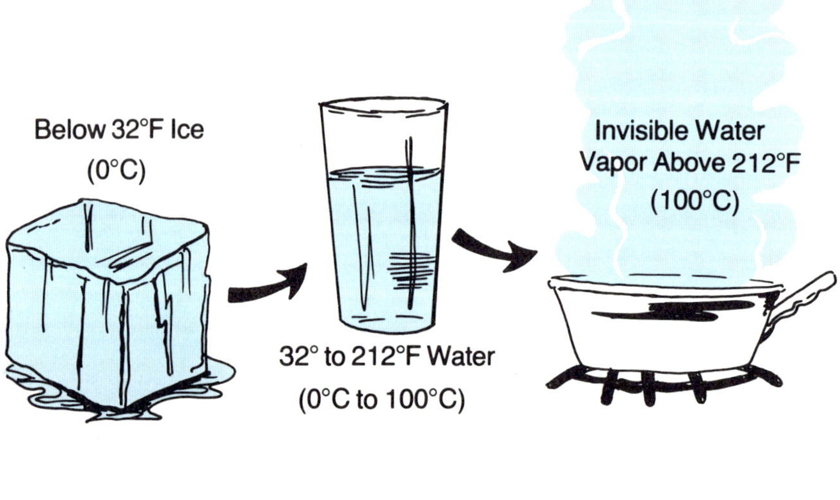

Water is for all practical purposes considered to be incompressible, and its weight varies at different temperatures. Water weighs the most (approximately 62.4 pounds per cubic foot [1 kilogram per liter]) close to its freezing point; it weighs the least (approximately 60 lb/ft^3 [0.96 kg/l]) close to its boiling point. When the weight of water is considered, one is actually speaking of its density or its weight per unit of volume; hence, the reference of pounds per cubic foot. For all general purposes, ordinary fresh water can be considered to weigh 62.5 lb/ft^3 (1kg/l).*

Heat Transfer and Heat Absorption

Three natural laws affect the use of water as an extinguishing agent and they are vitally important when the heat-absorbing ability of water is considered. These three natural laws are The Law of Heat Flow, The Law of Specific Heat and The Law of Latent Heat of Vaporization.

THE LAW OF HEAT FLOW

Heat tends to flow from a hot substance to a cold substance. This phenomenon established the ability of one substance to absorb heat from another substance. Fire can spread throughout a burning building by one or more of four methods commonly referred to as *conduction, radiation, convection* and *direct flame contact.*

Conduction is the transfer of heat from one body to another by direct contact through an intervening heat-conducting medium.

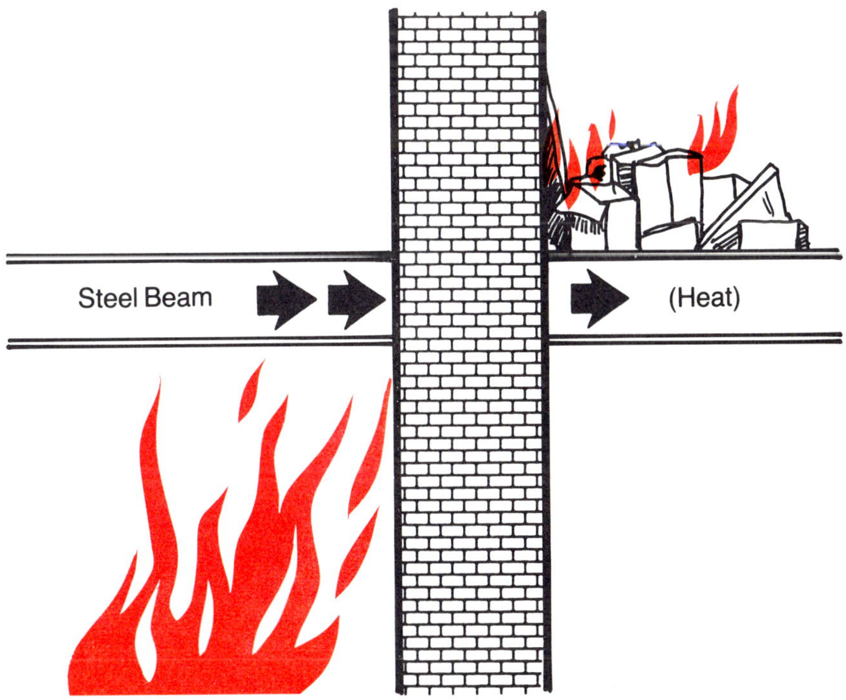

Figure 1.5 — Heat transfer by conduction.

*For an explanation of the metric system and its use in this manual see "THE METRIC SYSTEM," Chapter 3.

The amount of heat that will be transferred and its rate of travel depends upon the material through which the heat is passing.

Radiation is the transfer of heat through space from one body to another by heat rays. Radiated heat is not absorbed by the air to a great extent and heat will travel through space until it meets an opaque object.* As the object is exposed to heat radiation it will in return radiate heat from its surface. Radiated heat is one of the major sources of fire spread and its importance demands an immediate attack at points where radiation transfer is severe.

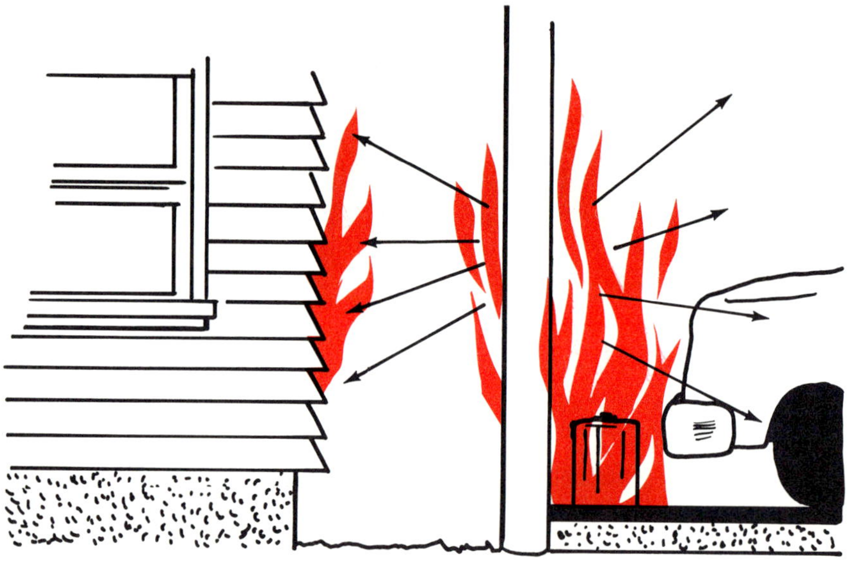

Figure 1.6 — Heat transfer by radiation.

Convection is the transfer of heat by the movement of air currents. Heated air in a building will expand and rise. For this reason, fire spread by convection is mostly in an upward direction, although air currents can carry heat in any direction, especially laterally. Convected heat currents are generally the cause of heat movement from floor-to-floor, room-to-room and area-to-area. The spread of fire through corridors, up stairwells and elevator shafts, between walls and through attics is mostly caused by convection of heat currents. The spread of fire by convection probably has more influence upon the positions for fire attack than either conduction or radiation.

Fire also spreads along and through a flammable material by direct flame contact. When a substance is heated to a point where flammable vapors are given off, these vapors may be ignited. If this substance is in contact with a burning material, the flame from the burning substance will heat the adjoining substance to a temperature where it may ignite and burn.

---

*"Opaque" in this sense means that the substance can stop radiated heat and light from passing through a given quantity of the substance.

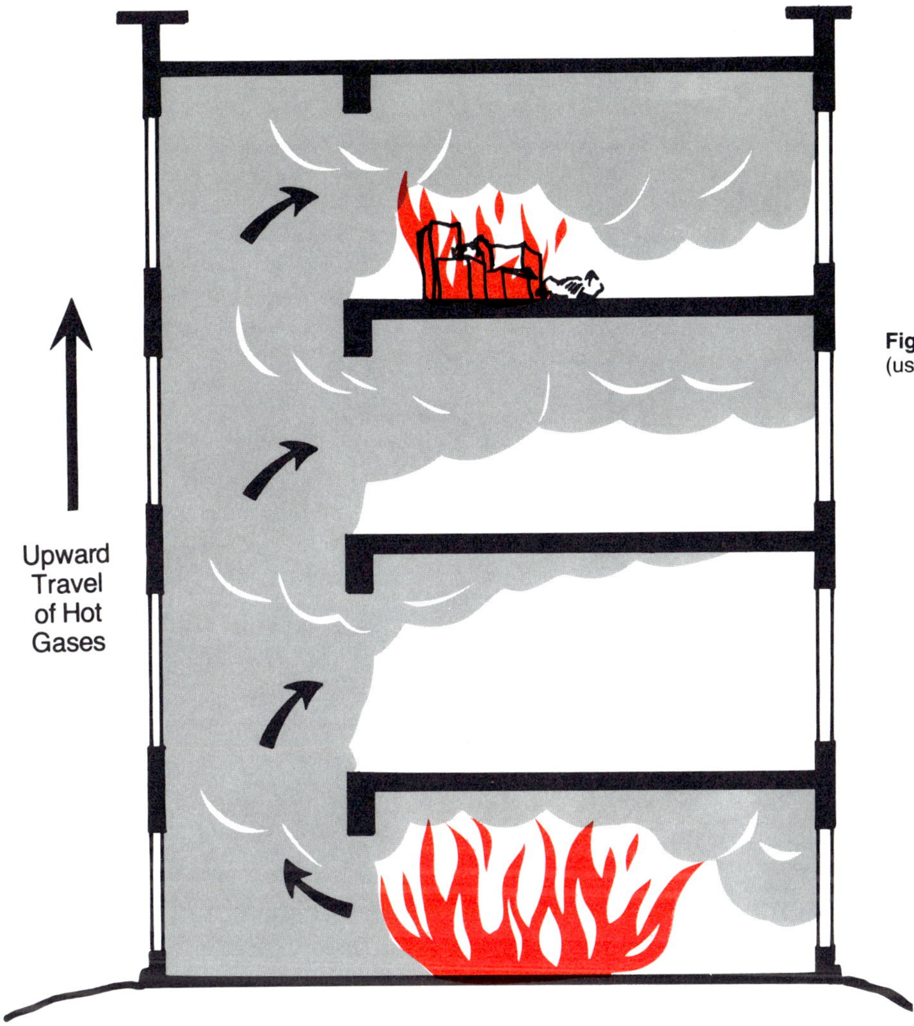

Figure 1.7 — Heat transfer by convection (usually upward).

Figure 1.8 — Heat transfer by direct flame contact.

Figure 1.9

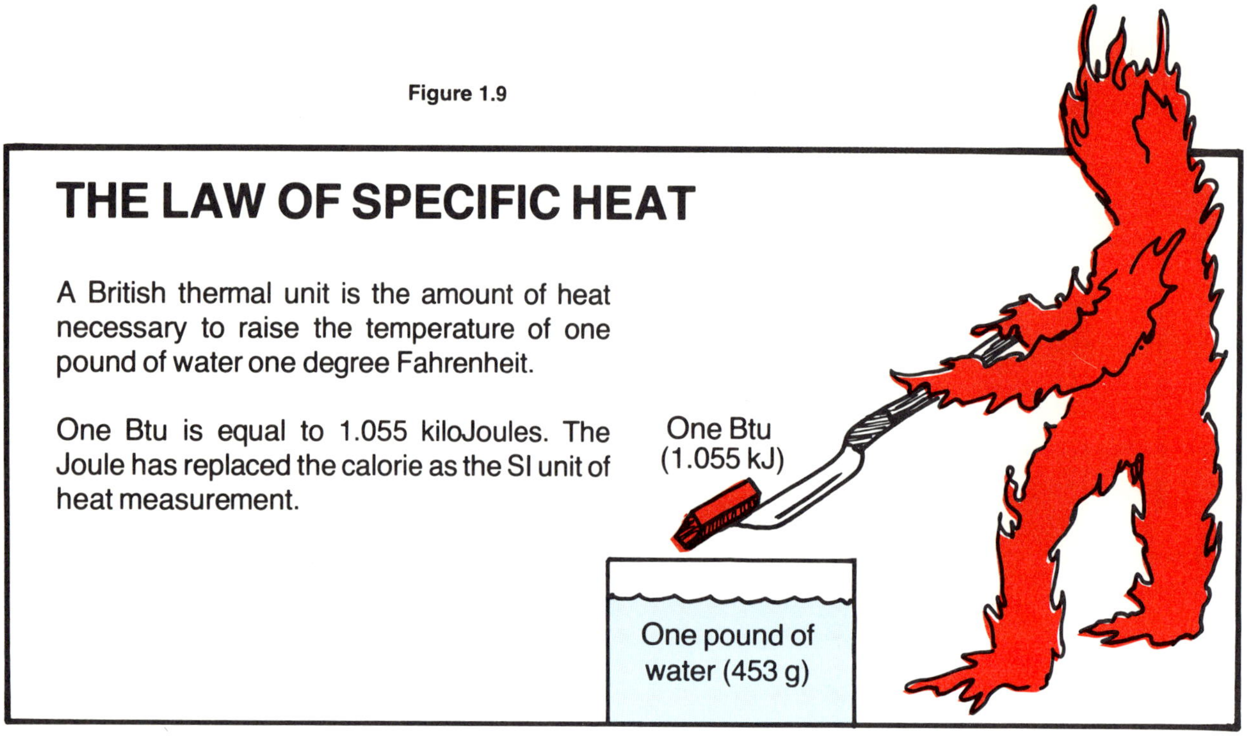

THE LAW OF SPECIFIC HEAT

Specific heat is a measure of the heat-absorbing capacity of a substance. Water is not only noncombustible; it also has the ability to absorb large amounts of heat. Amounts of heat transfer are measured in British thermal units or in Joules (1 Btu = 1.055 kJ). A Btu is the amount of heat required to raise the temperature of one pound of water 1°F. A Joule, also a unit of work, has taken the place of the calorie in the SI heat measurement (1 calorie = 4.19 Joules).

The specific heat of any substance is the ratio between the amount of heat necessary to raise the temperature of a specified quantity of a material and the amount of heat necessary to raise the temperature of an identical quantity of water by the same number of degrees. The specific heat of different substances varies. The following list, Table 1.1, of some fire extinguishing

TABLE 1.1
Specific Heat of Extinguishing Agents

Agent	Specific Heat
Water	1.00
Calcium chloride solution	0.70
Carbon dioxide (solid)	0.12
Carbon dioxide (gas)	0.19
Sodium bicarbonate	0.22

agents shows the specific heat comparison, by weight, with water. A comparison of the materials listed shows that water is the best material for absorbing heat.

By dividing the specific heat of water (1.00) by the specific heat of carbon dioxide gas (0.19), one finds that it takes more than fives times as much heat to raise the temperature of one pound of water 1°F as it takes for the same amount of carbon dioxide gas. The same comparisons may be made for the other materials listed. The result of these comparisons shows that water is better than any of these materials for absorbing heat from other substances.

THE LAW OF LATENT HEAT OF VAPORIZATION

The latent heat of vaporization is the quantity of heat absorbed by a substance when it changes from a liquid to a vapor. At sea level water boils at 212°F (100°C), at which time it vaporizes or turns to steam. This vaporization, however, does not happen at the instant the water reaches the boiling point. Each pound of water requires approximately 970 Btus (1,023 kJ) of additional heat to completely turn it into steam (see Figures 1.10 and 1.11). This characteristic is significant in fighting fire because the temperature of the water is not increased beyond 212°F (100°C) during the absorption of the 970 Btus for every pound of water. A gallon of water at 60°F (16°C) will, therefore, absorb more than 9,000 Btus (9,495 kJ) of heat if all of the water is converted into steam. The method by which this figure is derived is as follows:

$212° - 60° = 152°$;	specific heat = 1°F per pound
	1 gallon water = 8.33 pounds
Therefore:	$152(8.33) = 1{,}266$ Btu
And:	$970(8.33) = 8{,}080$ Btu
Then:	$1{,}266 + 8{,}080 = 9{,}346$ Btu

100 gallons of water absorbs 934,000 Btus

If the water (at 60°F, 16°C) from a 100 gpm (375 L/min) fog nozzle is projected into a highly heated area it has the ability to absorb approximately 934,000 Btus (985,400 kJ) of heat per minute. The quantity of heat produced when one pound of coal burns is approximately 10,000 Btus (10,500 kJ), a pound of wood produces about 8,000 Btus, and one gallon of gasoline around 130,000 Btus. The quantity of heat that may be produced by a combustible object depends upon the material of which it is composed. The rate at which the material gives off this heat depends on conditions such as its physical form, surface exposed and air or oxygen supply.

When a substance is heated or when it absorbs heat it may change either from a solid to a liquid or from a liquid to a vapor if enough heat is applied. The temperature at which a liquid absorbs enough heat to change it to a vapor is known as its boiling point. The ability of water to vaporize at the relatively low temperature of 212°F (100°C) makes it a very effective extinguishing agent.

The speed with which water absorbs heat is increased in proportion to the water surface exposed to the heat. For example, if a 1-inch (2.54 cm) cube of ice is dropped into a glass of water, it

Figure 1.10

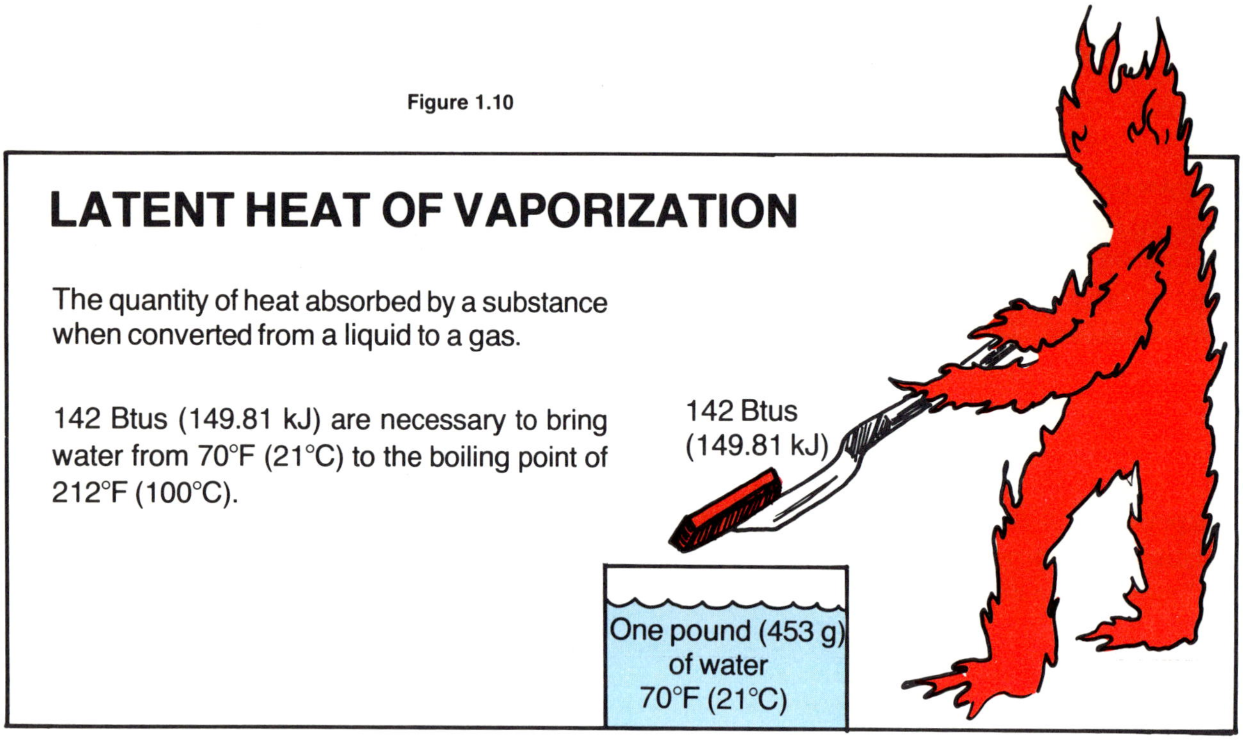

Figure 1.11

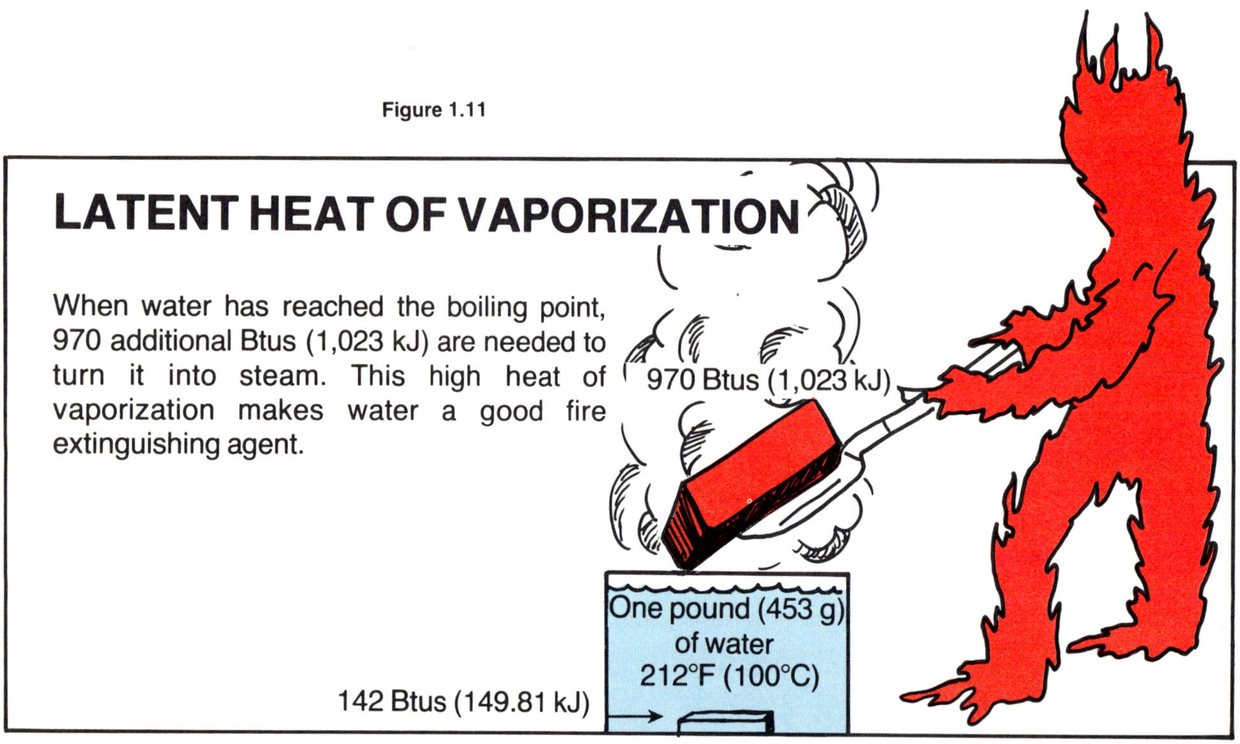

will take quite some time for the ice cube to absorb its capacity of heat because only 6 in^2 (38.7 cm^2) of the ice are exposed to the water. If, however, that cube of ice is divided into ⅛-inch (3.2 mm) cubes and they are dropped into the water, 48 in^2 (309.6 cm^2) of the ice are exposed to the water (see Figure 1.12). This principle also applies to water in a liquid state. If water is divided into many drops, the rate of heat absorption will be increased hundreds of times. A fog stream's finely divided particles of water will absorb heat from other materials more rapidly than if the water were applied as a solid stream. Specific heat figures are also significant in fire protection since they indicate the quantity of heat that must be removed to cool a hot substance from a dangerous to a safe temperature.

Figure 1.12

HEAT ABSORPTION

A 1-inch cube of ice exposes 6 square inches of surface area that can absorb heat.

The same cube cut into ⅛-inch cubes exposes 480 squares inches of surface and can absorb much more heat and do so more quickly.

One cm^3 of ice exposes 6 cm^2 of surface area that can absorb heat.

The same cube cut into millimeter cubes exposes 600 cm^2 that can absorb much more heat and do so more quickly.

Expansion of Water When Converted into Steam

Another way in which water is an aid to fire fighting is its expansion when converted into steam. When water changes into steam it expands. The amount of expansion varies with the temperature of the fire area. At 212°F (100°C) a cubic foot (28.3 cm^3) of water expands approximately 1,700 times its original volume. The greater the temperature, the higher the amount of expansion.

For example, at 500°F (260°C) the expansion ratio is approximately 2,400 times and at 1,200°F (649°C) the ratio is approximately 4,200 times. Table 1.2 gives the approximate volume that steam from one gallon (3.785 l) of water will occupy at different temperatures under normal atmospheric pressure (90 percent efficiency).

TABLE 1.2
Volume Occupied by Water Turned into Steam

Temperature (°F)	Temperature (°C)	One U.S. Gallon (ft^3)	One Liter (m^3)
212	100	225	6.35
300	150	250	7.00
400	200	275	7.80
500	260	300	8.50
600	315	325	9.20
700	370	350	9.90
800	425	400	11.30
900	480	450	12.75
1,000	540	500	14.00
1,200	650	550	15.50

It might be well to visualize a nozzle discharging 75 gallons (280 l) of water fog into a heated area of approximately 212°F (100°C) and the water being converted into steam. During this operation, 10 ft^3 (0.28 m^3) of water will have been vaporized and expanded to approximately 17,000 ft^3 (481 m^3) of steam. This is enough steam to fill a room approximately 10 ft (3 m) high, 25 ft (8 m) wide and 68 ft (21 m) long, as shown in Figure 1.13. In extremely hot atmospheres steam will further expand to greater volumes. Steam expansion is rapid. If the room is already full of smoke and gases the steam that is generated will displace these gases. As the room cools, the steam condenses and allows the room to refill with cooler air. The steam produced can also be an aid in fire extinguishment by smothering. Smothering is accomplished when the expansion of steam reduces the oxygen in a confined space.

Figure 1.13 — Water converted to steam expands 1,700 times. This expansion process absorbs heat and forces hot air and combustion gases out of a confined space.

Extinguishing Advantages of Water

Water may be used as a smothering agent when it will float on liquids, such as carbon disulfide, that are heavier than water. If the material is water soluble, such as alcohol, the smothering action will not take place. Water may also cause a smothering action on fires by the formation of an emulsion over the surface of certain flammable liquids.

Summing up the extinguishing properties of water, several advantages that are extremely valuable from the fire extinguishing point of view can be listed. The ultimate objective when using water to extinguish fire should be to take advantage of the following:

Water absorbs more heat than do other agents

- Water has a greater heat-absorbing capacity than any other common extinguishing agent.
- A relatively large amount of heat is required to change water into steam.
- The greater the surface area of the water that is exposed the more rapidly heat will be absorbed.
- Water converted into steam occupies several hundred times its original volume.

Disadvantages of Water as an Extinguishing Agent

Some disadvantages of water as an extinguishing agent include its surface tension, reactivity with certain materials, its low opacity and reflectivity, freezing temperature, low viscosity and conductivity.

Water has a considerable amount of surface tension and this property limits its soaking or penetrating ability into such things as baled or upholstered materials. (There are wetting agents that when mixed with water will reduce the surface tension of water and increase its penetrating ability.)

Water should not be used on some materials

Water will cause violent reactions when applied to some materials and chemicals. This reaction can sometimes occur even when the material is not burning. Water should not be used on such substances. NFPA's *Fire Protection Guide on Hazardous Materials* should be consulted for the proper extinguishing agents and methods for fires involving these substances.

Water will freeze at 32°F (0°C). This characteristic is a disadvantage since it can present a hazard to firefighters by coating equipment, roofs and other surfaces with ice. Also, ice forming in and on equipment may cause malfunction of the equipment.

Water has low viscosity (flows easily) — it does not readily adhere to vertical surfaces. This can be a disadvantage when exposures need protecting or when fire involves vertical or rounded surfaces.

Water that has been distilled of all impurities does not conduct electricity, but the water that a firefighter uses contains chemicals, minerals and organisms and will conduct electricity.*

PRINCIPLES OF PRESSURE

The word *pressure* has a variety of meanings. Ordinarily one thinks of pressure as an application of force on one thing by another thing. Pressure further suggests such synonyms as compression, pushing or squeezing, and has often been used to express a burden of physical or mental stress.

Pressure can easily be confused with *force*. Force is the simple measure of weight, usually expressed in pounds or kilograms. Pressure is *force per unit area*. The measurement of force by weight is directly related to the force of gravity, which is the amount of attraction the earth has for all bodies. If a given number of objects of the same size and weight are placed upon a flat surface they will each exert an equal force on that surface. For example, three square containers of equal size (1 × 1 × 1 foot [0.30 × 0.30 × 0.30 meter]) containing one cubic foot of water (28.3 dm^3) and weighing 62.5 pounds (28.35 kg) each are placed as illustrated in Figure 1.14. Each container exerts a force of about 62.5 pounds per square foot (psf) with a total of about 187.5 pounds (85 kg) of force over three square feet (0.8 m^2) of area. Pressure may be expressed in pounds per square foot (psf), pounds per square inch (psi) or in kiloPascals. If the containers are stacked on top of each other as shown in Figure 1.15, the total force (187.5 pounds; 85 kg) will remain the same, but the area of contact is reduced to one square foot (929.03 cm^2). The pressure then becomes 187.5 pounds per square foot.

Figure 1.14

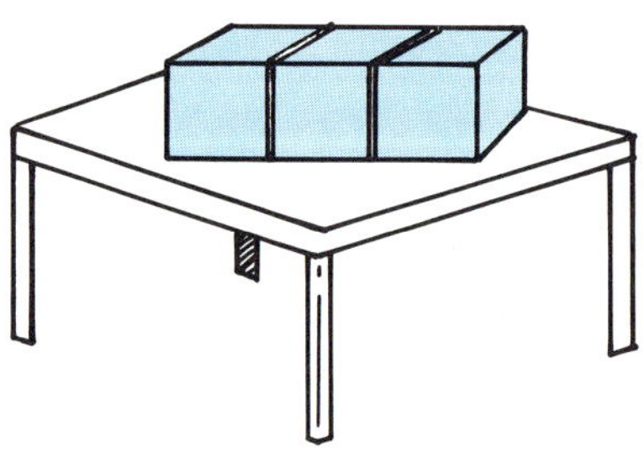

Figure 1.14 — Each container measures one cubic foot and weighs 62.5 pounds. Each container therefore exerts a pressure of 62.5 pounds per square foot.

Figure 1.15

Figure 1.15 — Each container measures one cubic foot and weighs 62.5 pounds. By being stacked, they exert a pressure of 187.5 pounds per square foot.

*Additional information concerning requirements for extinguishment of fires involving electrical equipment is discussed on page 35.

How Pounds Per Square Inch Are Determined

The weight of a cubic foot of water is approximately 62.5 pounds. Since one square foot contains 144 square inches, the weight of water in one square inch one foot high equals 62.5 divided by 144, or 0.434 pounds. A one-square-inch column of water one foot high will therefore exert a pressure at its base of 0.434 psi. The height required for a one-square-inch column of water to produce one psi at its base equals 1 divided by 0.434, or 2.304 feet; therefore, 2.304 feet of water will exert a pressure of one psi at its base, as shown in Figure 1.16.

Figure 1.16 — One cubic foot of water contains 144 cubic inches of water. The pressure exerted by a column of water one foot high is 0.434 pounds per square inch. To exert a pressure of one pound per square inch, a column of water must be 2.304 feet high (about 27½ inches).

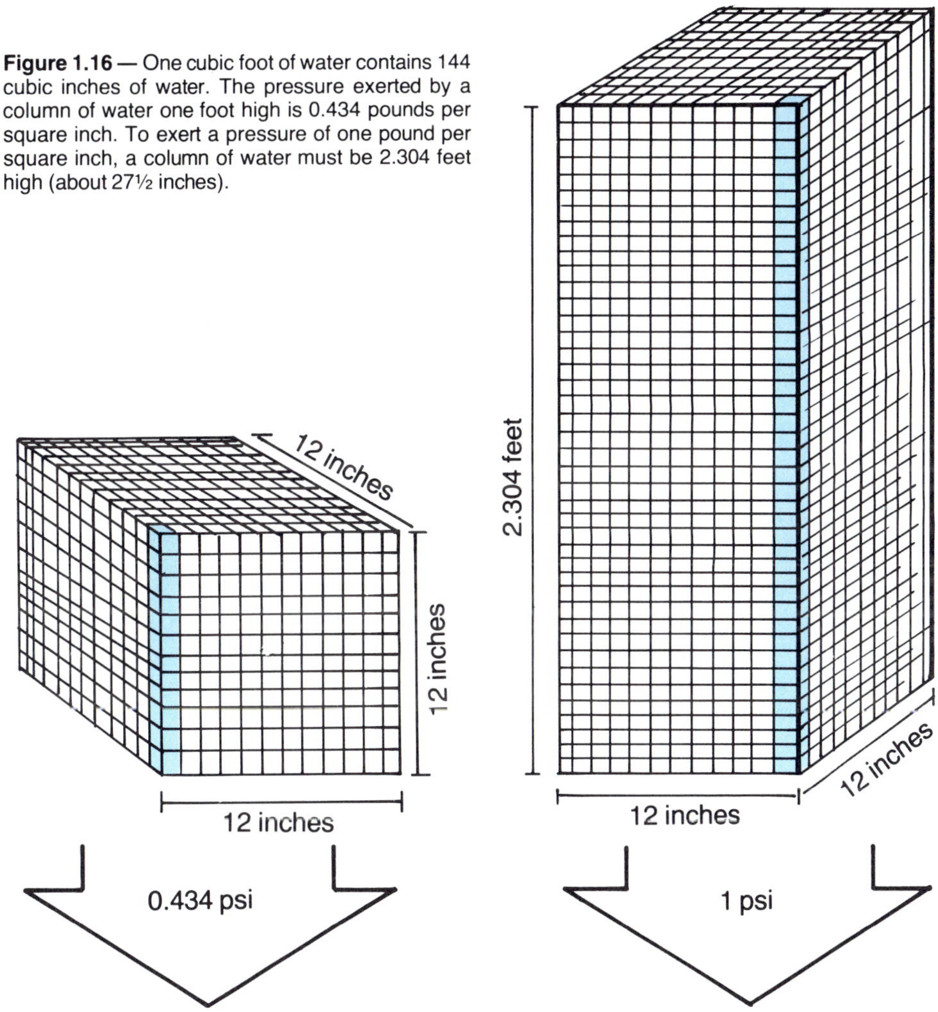

How Pressures Act on Fluids

The speed with which a fluid travels through hose is developed by pressure upon that fluid. The speed of travel is often referred to as *velocity,* and the kind of pressure should be identified because the word *pressure* in connection with fluids has a very broad meaning. There are six basic principles that determine the action of pressure upon fluids and they should be clearly understood before the kinds of pressures are studied.

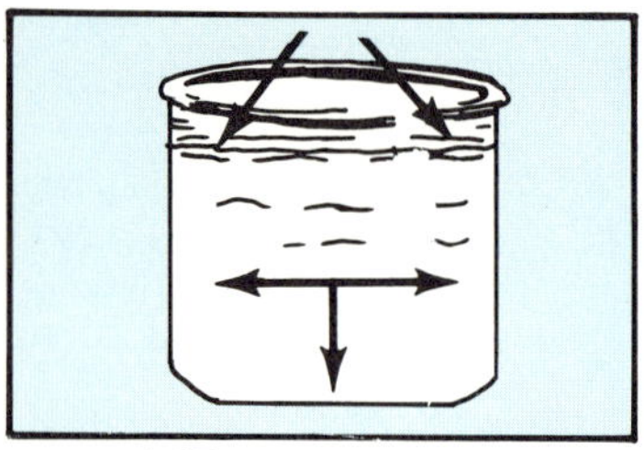
Figure 1.17

Figure 1.18

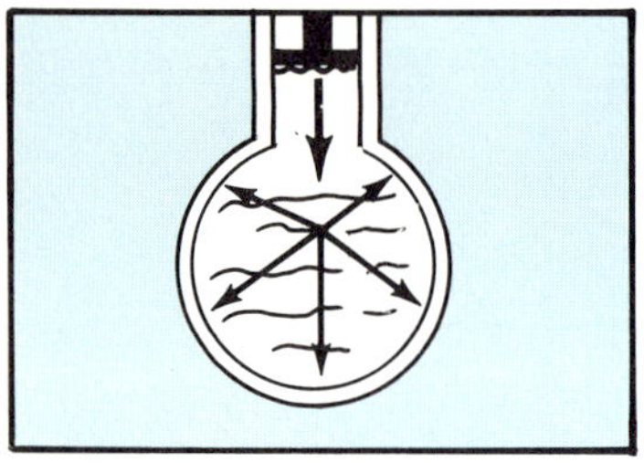
Figure 1.19

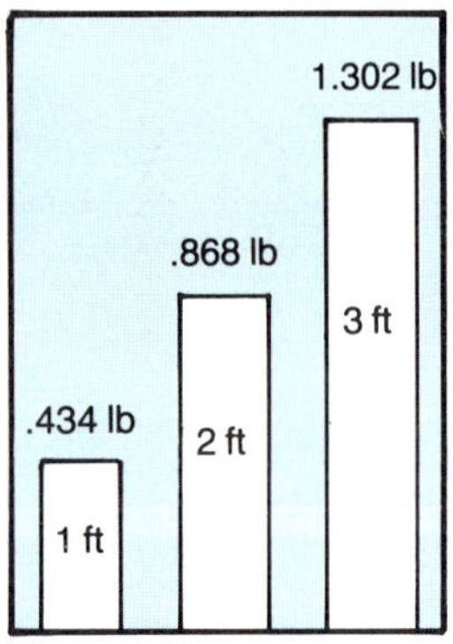

Figure 1.20

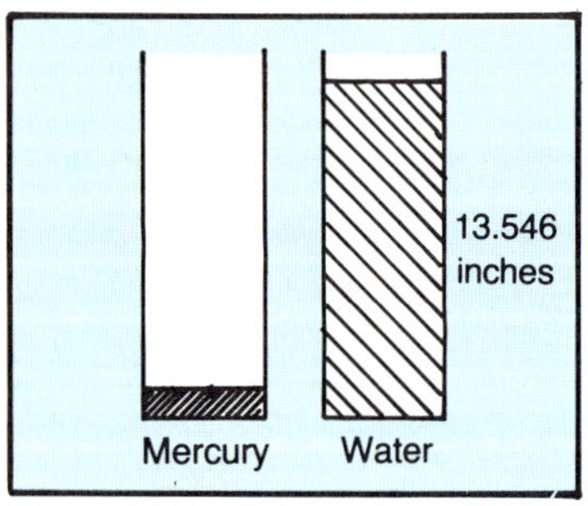

Figure 1.21

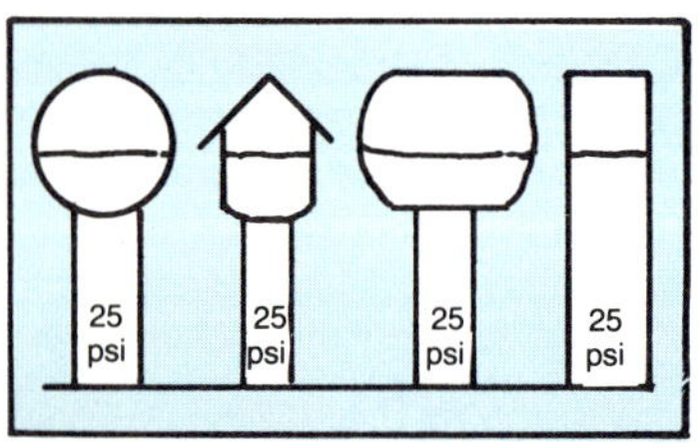

Figure 1.22

THE FIRST PRINCIPLE

Fluid pressure is perpendicular to any surface on which it acts. This principle is illustrated by a vessel having flat sides and containing water. The pressure exerted by the weight of the water, is perpendicular to the walls of the container. If this pressure is exerted in any other direction, as indicated by the slanting arrows, the water would start moving downward along the sides and rising in the center.

THE SECOND PRINCIPLE

Fluid pressure at a point in a fluid at rest is of the same intensity in all directions; or, it could be stated that fluid pressure at a point in a fluid at rest has no direction.

THE THIRD PRINCIPLE

Pressure applied to a confined fluid from without is transmitted equally in all directions. This principle is illustrated by a hollow sphere to which a water pump is attached. A series of gauges is set into the sphere around its circumference. With the sphere filled with water and pressure applied by the pump all gauges register the same, provided they are on the same grade line with no change in elevation.

THE FOURTH PRINCIPLE

The pressure of a liquid in an open vessel is proportional to its depth. This principle is illustrated by three vertical containers each being one square inch in cross-sectional area. The depth of the water is one foot in the first container, two feet in the second and three feet in the third. The pressure at the bottom of the second container is twice that of the first and the pressure at the bottom of the third container is three times that of the first. Thus, the pressure of a liquid in an open container is proportional to its depth.

THE FIFTH PRINCIPLE

The pressure of a liquid in an open vessel is proportional to the density of the liquid. This principle is illustrated by two containers, one holding mercury one inch in depth, the other holding water 13.55 inches in depth. The pressure at the bottom of each container is approximately the same since mercury is approximately 13.55 times denser than water. Hence, the pressure of a liquid in an open vessel is proportional to the density of the liquid.

THE SIXTH PRINCIPLE

The pressure of a liquid on the bottom of a vessel is independent of the shape of the vessel. This principle is illustrated by a number of containers of various shapes, each having the same cross-sectional area at the bottom and having the same height.

The pressure exerted at the bottom of each will be exactly the same if each vessel is filled with the same kind of liquid. The pressure of the liquid on the bottom of a vessel is, therefore, independent of the shape of the vessel.

Kinds of Pressure on Fluids

ATMOSPHERIC PRESSURE

The atmosphere surrounding the earth has depth and density and exerts pressure upon everything on earth. Atmospheric pressure is greatest at low altitudes; consequently, its pressure at sea level is used as a standard. At sea level the atmosphere exerts a pressure of 14.7 psi (100 kPa).

A common method of measuring atmospheric pressure is by comparing the weight of the atmosphere with the weight of a column of mercury: the greater the atmospheric pressure, the taller the column of mercury. The pressure of 1 psi will make the column of mercury about 2.304 inches tall. At sea level, then, the column of mercury will be 2.304 × 14.7, or 29.9 inches tall (see Figure 1.23).

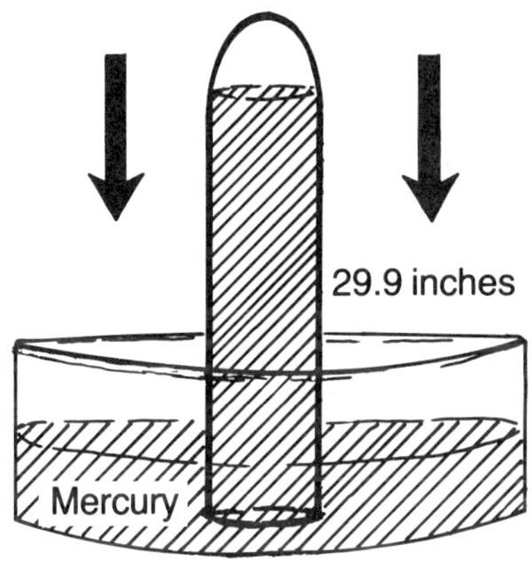

Figure 1.23 — A simple barometer. Atmospheric pressure forces the mercury into the tube. A pressure of 1 psi will cause the mercury column to rise 2.304 inches.

The readings of most pressure gauges are psi *in addition to* the existing atmospheric pressure. For example, a gauge reading 10 psi at sea level is actually indicating 24.7 psi (14.7 + 10). Engineers make the distinction between such a gauge reading and actual atmospheric pressure by writing *psig,* which means "pounds per square inch gauge." The notation for actual atmospheric pressure is *psia,* meaning "pounds per square inch absolute" (the psi above a perfect vacuum, absolute zero). Throughout this manual *psi* means *psig.*

When a gauge reads −5 psig it is actually reading 5 psi less than the existing atmospheric pressure (at sea level, 14.7 − 5, or 9.7 psia). Any pressure less than atmospheric pressure is called *vacuum.* Absolute zero pressure is called a *perfect vacuum.*

HEAD PRESSURE

Head, in the fire service, refers to the height of a water supply above the discharge orifice. In Figure 1.24 the head is 100 — the top of the water supply is 100 feet above the hydrant discharge opening. To convert head in feet to head in pressure, all that has to be done is to divide the number of feet by 2.304 (the number of feet that 1 psi will raise a column of water). The water source in Figure 1.24 has a head pressure of 43.4 psi.

STATIC PRESSURE

The word *static* means at rest or without motion. Pressure on water may be produced by an elevated water supply, by atmospheric pressure or by a force pump. If the water is not moving the pressure exerted is static. In water distribution systems there is

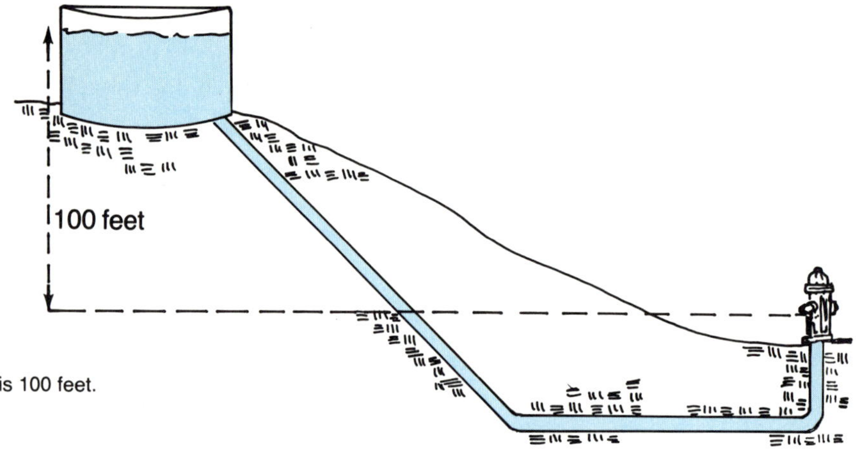

Figure 1.24 — The head in this illustration is 100 feet. The head pressure is 43.4 psi.

TABLE 1.3
Head in Feet and Head Pressure

Feet of Head	Pounds per Square Inch	Pounds per Square Inch	Feet of Head
5	2.17	5	11.50
10	4.33	10	23.00
15	6.50	15	34.60
20	8.66	20	46.20
25	10.83	25	57.70
30	12.99	30	69.30
35	15.16	35	80.80
40	17.32	40	92.30
45	19.49	45	103.90
50	21.65	50	115.40
60	26.09	60	138.50
70	30.30	70	161.60
80	34.60	80	184.70
90	39.00	90	207.80
100	43.30	100	230.90
110	47.60	110	253.90
120	52.00	120	277.00
130	56.30	130	300.10
140	60.60	140	323.20
150	65.00	150	346.30
160	69.20	160	369.40
180	78.00	170	392.50
200	86.60	180	415.60
250	108.20	190	438.90
300	129.90	200	461.70
350	151.50	225	519.50
400	173.20	250	577.20
500	216.50	275	643.00
600	259.80	300	692.70
700	303.10	325	750.40
800	346.40	350	808.10
900	389.70	400	922.60
1,000	433.00	500	1,154.50

Source: Waterous

always some flow in the pipes because of normal domestic or industrial needs. A truly static pressure is therefore seldom found in municipal water systems. From a practical viewpoint, however, the pressure normally found in a water system before water flows from a hydrant is considered to be static pressure. A waterflow definition of static pressure could be as follows: Static pressure is stored potential energy that is available to force water through pipe, fittings, fire hose and adapters.

NORMAL OPERATING PRESSURE

Normal operating pressure is the pressure when some water is flowing. The difference between static and normal operating pressure is the friction caused by water flowing through the various pipes, valves and fittings in the system. As soon as water starts to flow through a distribution system static pressure no longer exists. Water flows through a distribution system to provide a water supply to consumers and this flow will be increased if there are leaks in the system. The demands for water consumption continuously change during the day and night and the flow of water will therefore increase or decrease in the system. A waterflow definition of normal operating pressure can be as follows: Normal operating pressure is that pressure which is normally found on a water distribution system during normal consumption demands.

Residual pressure varies with amount of water used

RESIDUAL PRESSURE

The word *residual* means a remainder or that which is left. As an example, during a fire flow test the term *residual* represents the pressure that is left in a distribution system within the vicinity of one or more flowing hydrants. Residual pressure in a water distribution system will vary depending upon the amount of water that may be flowing from one or more hydrants, water consumption demands and the size of the pipe. One point that must be remembered is that residual pressure must be identified at the location where the reading is taken. A waterflow definition of residual pressure could be as follows: Residual pressure is that part of the total available pressure that is not used to overcome friction or gravity while forcing water through pipe, fittings, fire hose and adapters.

FLOW PRESSURE

The rate of flow (velocity) of the water coming from a discharge opening produces a force called *flow pressure* or *velocity pressure.* Since a stream of water emerging from a discharge opening is not encased within a tube it exerts pressure in a forward direction but does not exert sideways pressure. The forward velocity of flow pressure can be measured by using a Pitot tube and gauge. If the size of the opening is known the flow pressure can be used to calculate the quantity of water flowing in gallons per minute (gpm) (liters per minute [L/min]). An example of flow pressure is one in which the forward velocity of a water stream exerts a pressure that can be read on a gauge. A waterflow definition of flow pressure could be as follows: Flow pressure is the forward velocity pressure at a discharge opening, indicated by a Pitot tube and gauge while water is flowing.

Pressure "Loss" and "Gain": Elevation, Altitude

Although *elevation* and *altitude* are often used interchangeably in everyday life, the fire service makes a distinction between the two. Elevation refers to the position of something above or below grade level or, more usually, the centerline of the pump. Altitude is the position of something above or below sea level. Both have importance in the production of fire streams.

Because of gravity, when a nozzle is above the pump there is a pressure "loss," when below the pump there is a pressure "gain." Both "loss" and "gain" are about 0.5 psi per foot, or about 5 psi per conventional building story (see Figures 1.25 and 1.26). The pressure "loss" from elevation is also called *back pressure*.

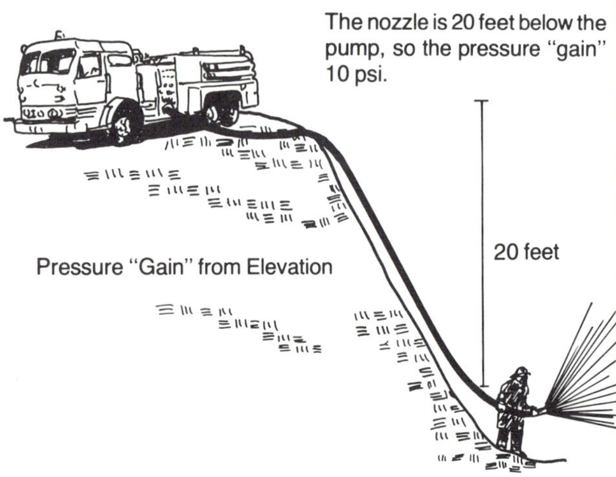

Figure 1.25

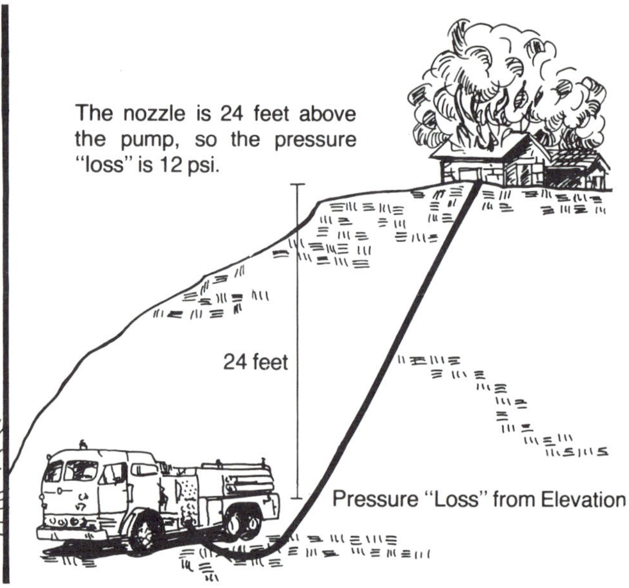

Figure 1.26

EXAMPLE (Figure 1.27)

The pumper discharge pressure reads 100 psi. The nozzle is going to be elevated to the fourth story of a building. What pump discharge pressure is necessary to maintain the fire stream?

There is a pressure "loss" of about 5 psi per story above the first, so the additional pump discharge pressure necessary is 3(5) = 15 psi. To maintain the fire stream the pump discharge pressure will have to be 115 psi.

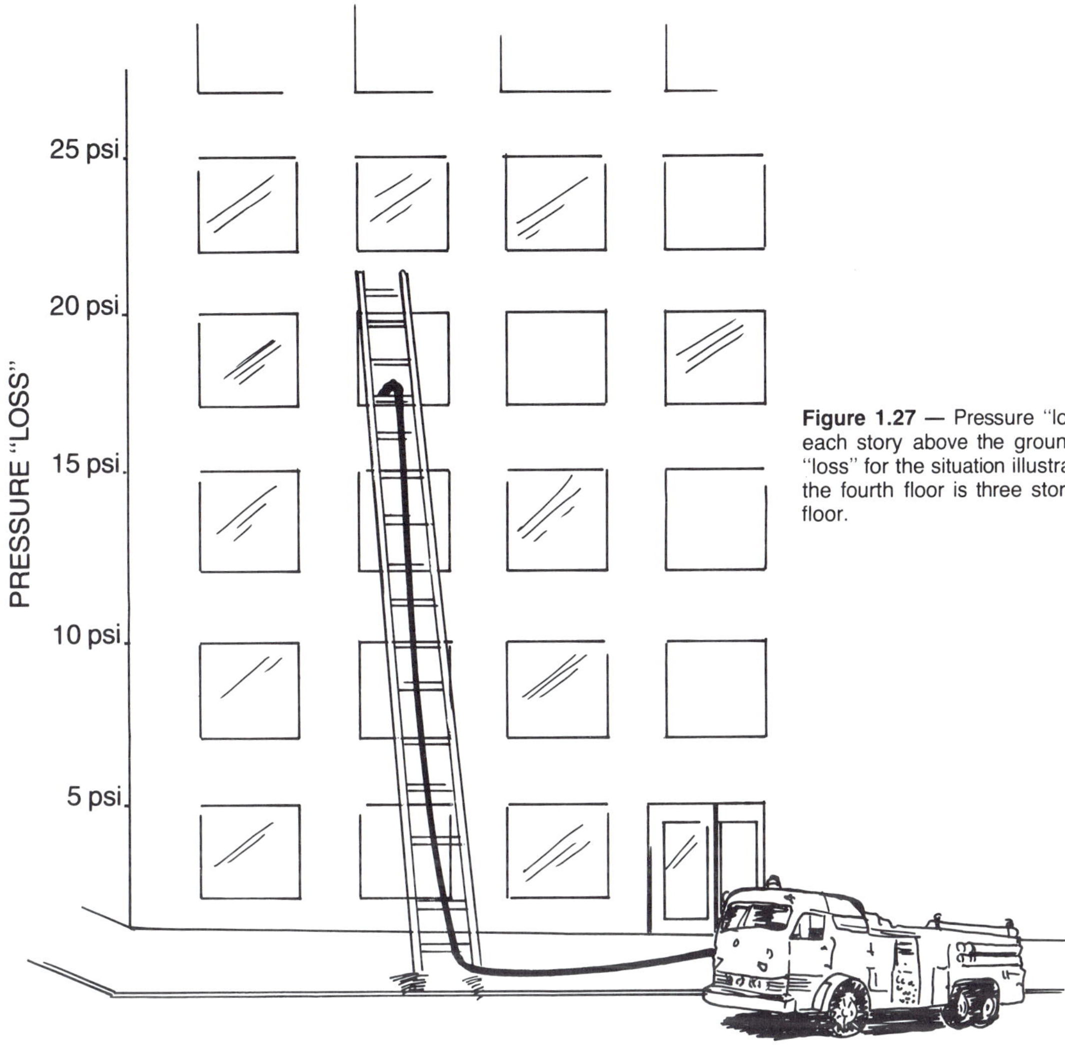

Figure 1.27 — Pressure "loss" is about 5 psi for each story above the ground floor. The pressure "loss" for the situation illustrated is about 15 psi — the fourth floor is three stories above the ground floor.

TABLE 1.4
Pressure and Lift at Various Altitudes

Altitude feet (meters)		Atmospheric Pressure psi (kPa)		Maximum Theoretical Lift feet (meters)	
Sea level		14.7	(100.00)	33.9	(10.33)
1,000	(300)	14.2	(97.90)	32.8	(9.99)
2,000	(600)	13.7	(94.46)	31.6	(9.63)
3,000	(900)	13.2	(91.01)	30.5	(9.29)
4,000	(1,200)	12.7	(87.56)	29.3	(8.93)
5,000	(1,500)	12.2	(84.11)	28.2	(8.59)
6,000	(1,800)	11.7	(80.67)	27.1	(8.26)
7,000	(2,100)	11.3	(77.91)	26.0	(7.92)

Altitude affects all aspects of the production of fire streams because with increasing height above sea level the atmospheric pressure drops. This pressure drop is of little consequence up to about 1,000 feet, but above that the lessened atmospheric pressure can be of concern. The pumpers of fire departments at high altitudes must work harder to produce the pressures required for effective fire streams, and their effective lift when drafting is reduced because of the less dense atmosphere.

At elevations above sea level the atmospheric pressure decreases approximately 0.5 psi (3.5 kPa) for every 1,000 feet (300 m).

Water Hammer

Water hammer can cause great damage

When the flow of water through fire hose or pipe is suddenly stopped the resulting surge is referred to as *water hammer.* Water hammer can often be heard as a distinct sharp clank, very much like a hammer striking a pipe. This sudden stopping results in a change in the direction of energy and this energy is instantaneously multiplied many times. These excessive pressures can cause considerable damage to water mains, plumbing, fire hose, fire pumps and can possibly injure firefighters. Nozzle controls, hydrants, valves and hose clamps should be operated slowly to prevent water hammer.

Friction Loss

Water flowing through pipe or fire hose meets certain resistances or friction that must be overcome by pressure. This loss of pressure is usually called *friction loss* or *loss because of friction.* The only pressure available to overcome this resistance is the total pressure. A fire stream definition of friction loss could be as follows: Friction loss is that part of total pressure that is used to overcome friction while forcing water through pipe, fittings, fire hose and adapters. The difference in pressure on a hoseline between a nozzle and a pumper is a good example of friction loss.

Friction loss in a hoseline is caused by the movement of the water molecules against each other, rough linings in fire hose, crushed couplings, sharp bends, change in size of orifice by adapters and improper gasket size. Anything that affects the movement of the water may cause additional friction loss. Fire hose of good quality has a smoother inner surface and will cause less friction than will hose of a lower grade. Hoselines that are laid in zigzag patterns have greater loss than hose laid in a straight line. The friction loss in old hose may be as much as 50 percent greater than that in new hose.

There are four basic principles that govern friction loss in fire hose and pipes.

THE FIRST PRINCIPLE

All other conditions being the same, the loss by friction varies directly with the length of the hose or pipe. This principle can be illustrated by a hoseline 100 ft (30 m) long and another line 200 ft (60 m) long. A constant flow of 200 gpm (750 L/min) is maintained in each hose. The 100-foot (30 m) length has 10 psi (7.0 kPa) and the 200-foot (60 m) length has twice as much loss or 20 psi (14.0 kPa).

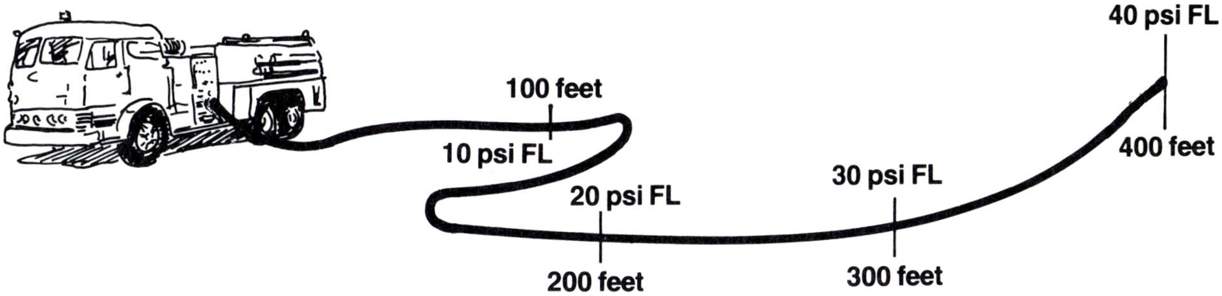

Figure 1.28 — Friction loss increases in a direct proportion to the length of the conduit if the conduit is of uniform diameter and smoothness.

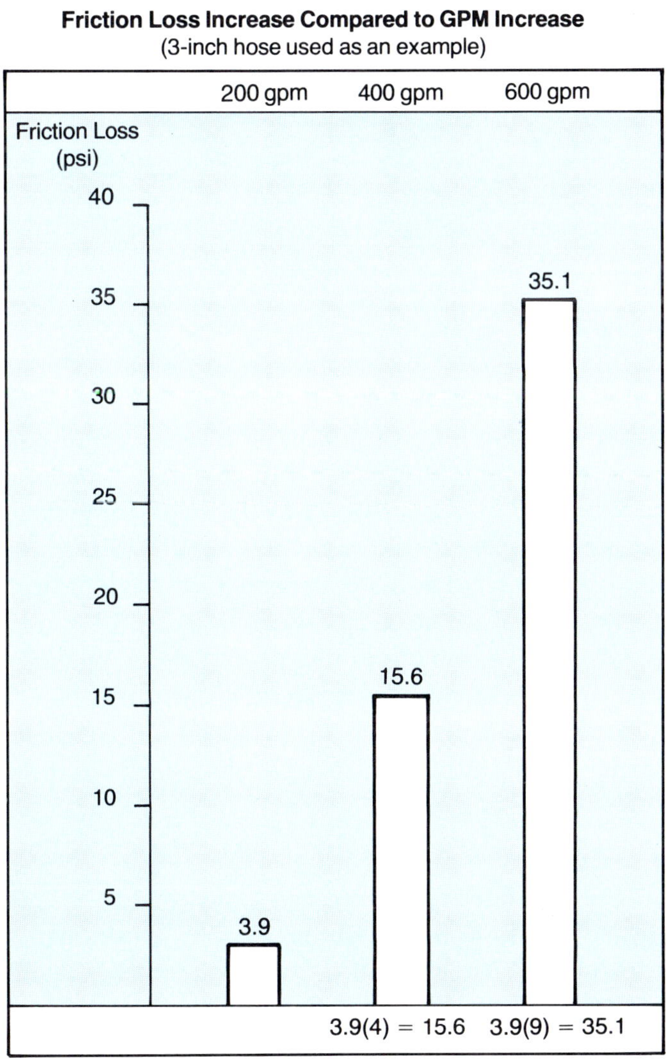

Figure 1.29 — An increase in velocity increases friction loss. The friction loss increase is much higher (not the same as) than the velocity increase.

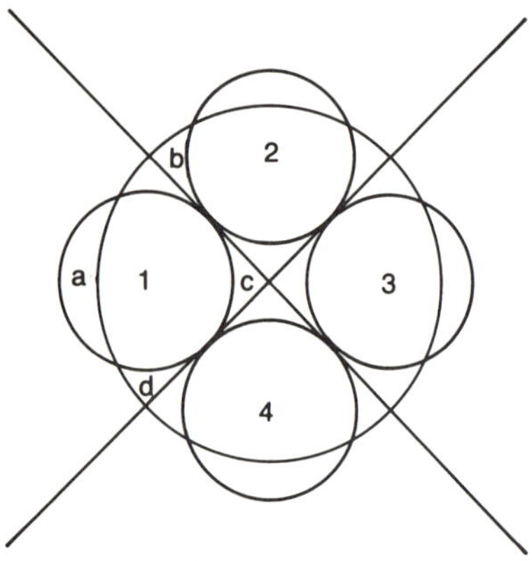

1 = 2= 3= 4
a = b + c + d

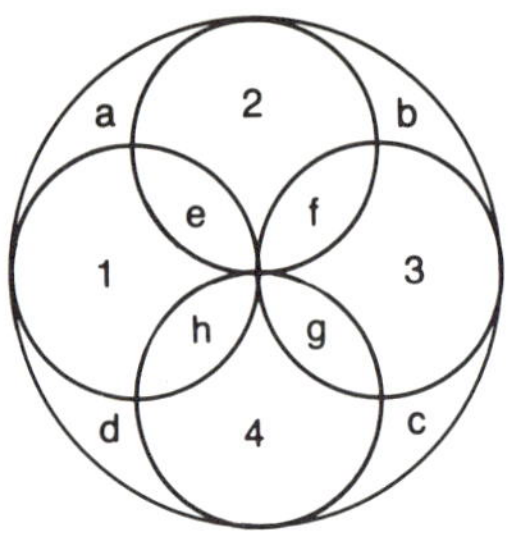

1 = 2 = 3 = 4
a = e, b = f, c = g, d = h

Figure 1.30

THE SECOND PRINCIPLE

When the same size of hose is used, the friction loss will vary approximately with the square of the increase in the velocity of the flow. This principle points out that friction loss develops much faster than the change in velocity. It must be remembered that the variation is only approximate to the square. The principle can be illustrated by three sections of 3-inch (75 mm) hose (Figure 1.29).

THE THIRD PRINCIPLE

For the same discharge, friction loss varies inversely as the fifth power of the diameter of the hose. This principle readily proves the advantage of larger size hose and can be illustrated by one hoseline 2½ inches (65 mm) in diameter and another hoseline 3 inches (75 mm) in diameter. The friction loss in the 3-inch (75 mm) line is:

$$\frac{2\tfrac{1}{2}^5}{3^5} = \frac{98}{243} = 0.4 \text{ that of the } 2\tfrac{1}{2}\text{-inch hose}$$

TABLE 1.5
Effect of Hose Diameter on Friction Loss

Diameter inches (mm)	Length feet (m)	Flow gpm (L/min)	Friction Loss psi (kPa)
2½ (65)	100 (30)	500 (1,900)	55 (380)
3 (75)	100 (30)	500 (1,900)	22 (150)

THE FOURTH PRINCIPLE

For a given velocity of flow, the friction loss in hose is approximately the same whatever the pressure on the water may be. This principle explains why friction loss will be the same when 250 gpm (940 L/min) is flowing through each of two hoselines, one with a pressure 130 psi (900 kPa), the other with 75 psi (500 kPa). If a given quantity of water, for example 100 gpm (375 L/min), must pass through a 3-inch (75 mm) hoseline in a certain length of time, the water must travel at a specified velocity (feet per second). If, however, this same gpm must pass through one 1½-inch (38 mm) hoseline, the velocity must be greatly increased. Four 1½-inch (38 mm) hoselines are required to flow 100 gpm (375 L/min) at a velocity equal to that required for the 3-inch (75 mm) hoseline.

---

*Because of the inherent tendency of some modern fire hose to stretch to a larger inside diameter under higher pressures, it must be kept in mind that this will decrease the velocity and therefore will decrease the friction loss. If in-line gauges are used for checking friction loss a difference may be noted.

TABLE 1.6
Effect of Pressure on Friction Loss

Diameter inches (mm)	Length feet (m)	Flow gpm (L/min)	Nozzle Pressure psi (kPa)	Engine Pressure psi (kPa)	Friction Loss psi (kPa)
2½ (65)	200 (60)	250 (950)	100 (700)	130 (900)	30 (200)
2½ (65)	200 (60)	250 (950)	45 (310)	75 (520)	30 (200)

As previously stated, one of the physical properties of water is that it is practically incompressible. This property means that the same volume of water supplied into a fire hose under pressure at one end will be discharged at the other end. The velocity for a given volume of water is determined by the size of the hose — and the smaller the hose, the greater the velocity must be to deliver the same volume.

A condition existing in a practical water system and a fire hose layout is shown in Figure 1.31. An elevated tank is filled with water to a height of 150 feet (45 m). At the bottom of the tank are the pipe connections to the fire hydrant. From the hydrant 300 feet (90 m) of 2½-inch (65 mm) hose is laid along the street with a valve on the end. Imagine a glass tube connected to the hose every 100 feet (30 m) standing upright 150 feet (45 m) to the same height as the elevated tank. With the valve closed the water in all the tubes would stand at line A, the same level as the water in the tank. This line indicates the static pressure. When the valve on the nozzle end is opened, a fair flow at a low pressure is attained. If the opening had been made directly at the hydrant, the flow would

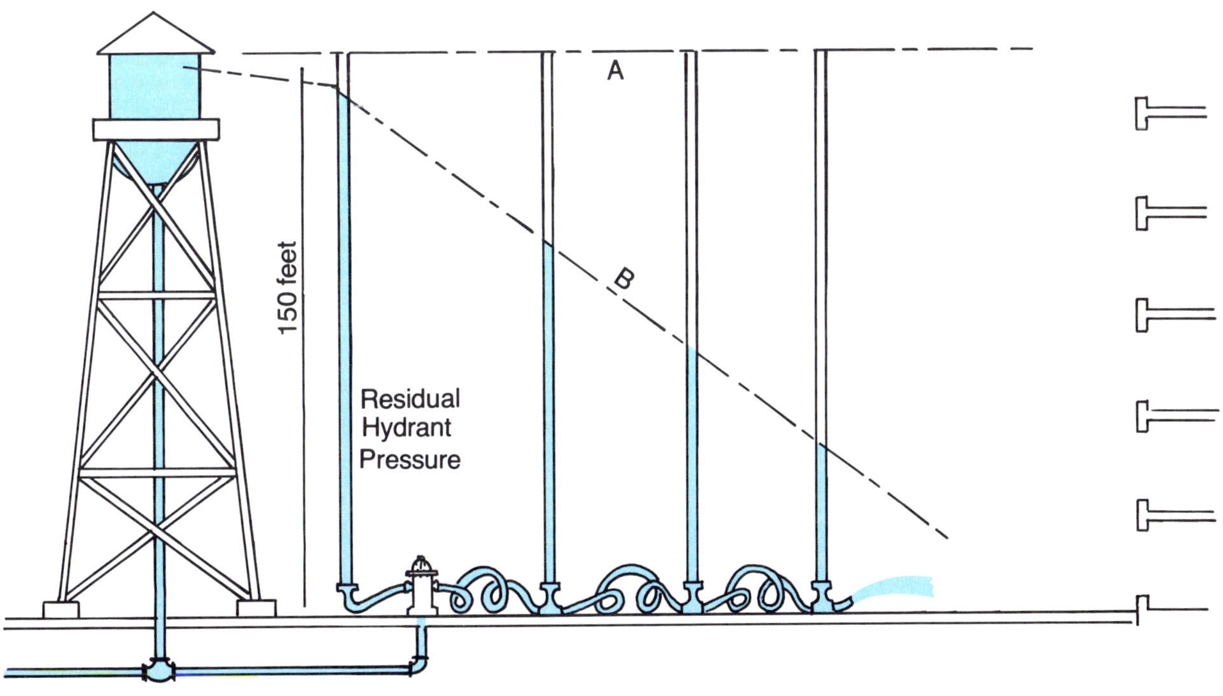

Figure 1.31 — Line A indicates static pressure. The distance between lines A and B indicates the pressure lost to friction. Note that there is no nozzle on the hose.

have been much greater at a higher pressure. In other words, the flow pressure is not as great at the end of the hoseline as it is at the hydrant. A look at the water level in the vertical tubes shows that instead of being up to that of the standpipe it is about as shown by line B. The difference in the water level of the tubes indicates the pressure that is used to overcome the friction loss in the sections of hose between the tubes. The loss of pressure at each 100 feet (30 m) interval and the reduced discharge indicate the friction loss in the line.

Since an open fire hose produces a stream that normally has no use in fire fighting the stream must be shaped by using a fire stream nozzle of some type. In Figure 1.32 a nozzle with a 1-inch (25 mm) solid stream tip has been used. With the nozzle closed the water level in the glass tube will be at line A, but with the nozzle open, it will drop to about line C. Notice that line C is considerably higher than line B in Figure 1.31 and that a fire stream with some reach has been produced. If the 1-inch (25 mm) tip is replaced with a 0.75-inch (20 mm) tip the water level in the glass tubes will be still higher and the reach will be greater, but the gallons per minute flow will be less. By decreasing the amount of water flowing the firefighter has reduced the speed of the water in the hose and consequently there is less friction loss.

Observe the height of the water in the first glass tube in Figures 1.31 and 1.32. The water level in these tubes indicates a good supply of residual pressure left in the water main. In the discussion of pressures it was established that residual pressure is that part of total pressure that is not used to overcome friction or gravity while forcing water through pipe, fittings, fire hose and adapters. Under these conditions it is advisable to place a pumper

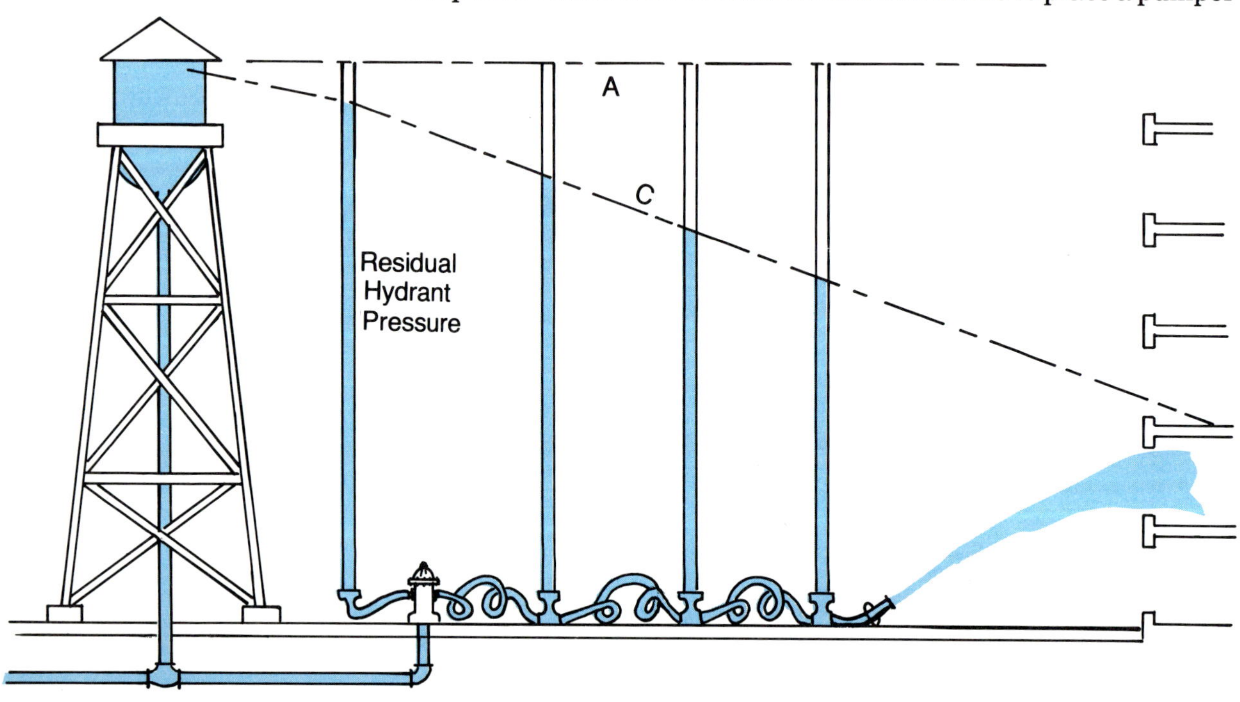

Figure 1.32 — A nozzle has been attached to the hose. This decreases the amount of water flowing per minute, which results in less friction loss (line C).

in the line at the hydrant. The use of a pump at this point provides additional force to increase the pressure in the hose to a point where effective fire streams may be produced. The addition of a pump also makes it possible to add hoselines even to the extent of providing master streams. This situation is illustrated in Figure 1.33. Here the water in the first tube has been lowered because of the additional flow of water.

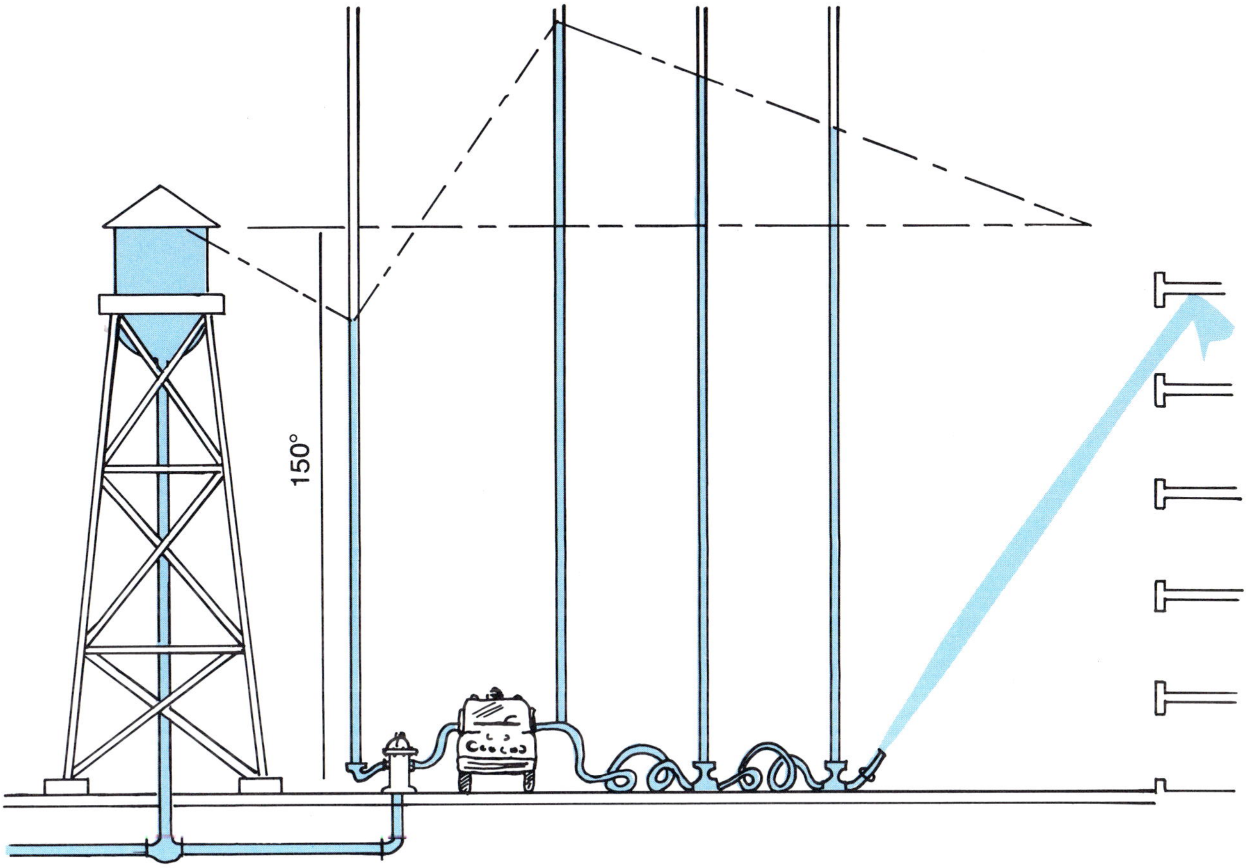

Figure 1.33 — A pumper is being used to increase the pressure on the water. Note the drop in residual pressure, and the rise in pressure indicated by the tubes to the right of the pumper.

Critical Velocity

One point that should be considered in applying pressure on water in hoselines is that there are practical limits to the velocity or speed at which a stream can travel. If the velocity is increased beyond these limits the friction becomes so great that the entire stream is agitated by resistance. This agitation causes a point of turbulence called *critical velocity*. Beyond this point it becomes necessary to parallel or siamese hoselines to increase the flow and reduce friction.

Reducing Friction Loss

Certain characteristics of hose layouts affect friction loss. To reduce pressure loss from friction as much as possible the following should be observed:

- Check for rough linings in old hose.

- Replace crushed couplings.
- Eliminate sharp bends.
- Use adapters only when necessary.
- Keep nozzles and valves fully open when possible.
- Use gaskets of the proper size.
- Use short lines as much as possible.
- When flow must be increased use larger hose or multiple lines.

THE USE OF WATER AS AN EXTINGUISHING AGENT

Structural Fires

Incipient fires need direct application

A fire in the incipient phase usually requires a direct application of water at the base of the fire to reduce the temperature of the fuel to below its ignition point. The amount of heat being generated will increase as the fire progresses. Although the flame temperature will be well over 1,000°F (538°C) the room temperature may only be slightly increased. The major consideration is reaching the fire.

During the free-burning phase the area will be extremely hot. A confined building fire in the second phase requires a reduction of atmospheric temperature to livable conditions so firefighters can enter and extinguish the main body of fire. Once the temperature is reduced firefighters may enter and extinguish any remaining fire by direct application, but positive-pressure self-contained breathing equipment will be required for safety.

During the smoldering phase the heated area will be filled with hot smoke and fire gases. If the fire gases and smoke were the only considerations firefighters could enter with positive-pressure self-contained breathing equipment — but the heat would be unbearable. Experiments and tests have shown that a building containing a third-phase fire may be made tenable by directing water fog into the most highly heated upper levels. The heat will convert the finely divided particles of water into steam, thereby reducing the temperature in the space. The expansion of steam will, further, force smoke and gases from the space. In this phase the heat must be reduced and the smoke and gases must be displaced before fresh air can be admitted, to prevent flashover or backdraft.

Open fires need large volumes of water

An open or unconfined building fire is one that has burned to the outside, releasing the heated smoke and gases. The entire building may be involved and these open fires require long-range operations. The intense radiated heat from an open fire makes it difficult to apply water directly at the base of the fire. A large volume of water is usually required to control and extinguish this type of fire, which may require straight or solid streams and long-reaching fog streams. This type of fire also requires careful consideration of exposures, with equal consideration of using fire streams to protect them.

Flammable Liquid Fires

Flammable liquids produce fuel vapors under varying circumstances according to the volatility of the liquid. The lowest temperature at which a flammable liquid will give off flammable vapors is known as its flash point. Gasoline, for example, gives off flammable vapors at a temperature as low as −45°F (−43°C), and kerosene starts giving off flammable vapors somewhere between 100°F (38°C) and 150°F (66°C). Their ignition temperatures range from 536°F (280°C) to 853°F (456°C) for gasoline and from 410°F (210°C) to 444°F (229°C) for kerosene. One can see that it may be difficult to adequately reduce with water fog the temperatures of these or other flammable liquids and that extinguishment of the fire may depend upon how the water fog is applied. Even though the fire is extinguished there is a constant danger of reignition if any materials near the liquid retain a temperature above the ignition temperature of the flammable liquid's vapors.

Kerosene and the heavier petroleum fuels do not produce sufficient fuel vapors for burning at ordinary atmospheric temperatures. A fire involving kerosene vapors may be extinguished with limited danger of reignition. For additional information concerning flash point, ignition temperature, and flammable limits of flammable liquids, together with the limitations of extinguishing agents, consult the latest edition of NFPA's *Fire Protection Handbook*.

Fires Involving Energized Electrical Equipment

A small fire in energized electrical equipment can usually be easily extinguished with the use of portable fire extinguishers. The extinguishing agent must not be a conductor of electricity. The use of water-base agents (including foam) may expose the extinguisher operator to electrical shock. For this reason Class C portable extinguishers are recommended.

Fire streams conduct electricity

The application of fire streams to electrical equipment should be avoided whenever possible because water is damaging to the windings, insulation and mechanisms. There are conditions, however, where the application of fog streams is highly desirable. For example, heat radiation from a nearby fire may ignite the crossarms and/or poles of overhead electrical wires, and if the fire is not extinguished the energized wires may fall to the ground. The fallen wires would be a serious threat to firefighters who are working in that area. Applications of water from fog nozzles to the crossarms and poles can be made with reasonable safety. Water from a fog nozzle is sufficiently broken to reduce the possibility of the stream becoming a conductor of electricity.

Firefighters should always consider every electrical wire to be energized and dangerous. Physical contact with wires or electrical equipment should be avoided and the distance between the nozzle and the electrical equipment should never be less than the distances given in Table 1.7. This table was taken from Insurance Services Office Special Interest Bulletin No. 91.

TABLE 1.7
Safe Distances for Using Water on Energized Electrical Equipment

	Solid Stream Nozzles			Fog Nozzles	
	1⅛-inch	1½-inch			
Volts	feet (m)	feet (m)		Volts	feet (cm)
1,100	6 (2)	9 (3)		15,000	½ (15)
2,200	11 (3)	16 (5)		25,000	¾ (20)
3,300	15 (5)	22 (7)		46,000	1¼ (40)
5,500	18 (5)	27 (8)		92,000	2½ (75)
6,600	19 (6)	29 (9)		138,000	3⅔ (120)
11,000	20 (6)	30 (9)		161,000	4⅓ (135)
22,000	25 (8)	33 (10)		230,000	6⅓ (200)
33,000	30 (9)	40 (12)		345,000	10 (320)

When it is *certain* that no current is flowing through electrical equipment the fire is then considered to be a Class A or Class B fire, depending upon the materials involved, and the appropriate Class A or Class B extinguishing agents may be used.

Table 1.7 applies to fresh water. Sea or salty water or water-base extinguishing agents may have such high electrical conductivity that no rule can be applied for a reasonable safe distance. If it becomes necessary to use water known to have a high electrical conductivity, a broken stream should be used and then with extreme caution. The distances given in the table are minimum and are considered safe under normal circumstances. A study of Table 1.7 will indicate that when a firefighter either intentionally or unintentionally directs water onto a charged conductor the fog nozzle is safest.

Fires Involving Combustible Metals

Water should not be used for fires in or near combustible metals such as titanium, zirconium, lithium, potassium and sodium. These metals react violently with water when they are involved in a fire. Some of them will react even when there is no fire (potassium and sodium, for instance, release the hydrogen from the water). Most such fires will occur in industrial occupancies, where special extinguishing agents will be readily available. If a combustible metal fire occurs on the highway or railway,

TABLE 1.8
Current Feedback from Live Grid through Fire Stream to Firefighter

Nozzle	Nozzle Pressure psi (kPa)	Flow gpm (L/min)	Grid Voltage	Ground Voltage	Milliamps Conducted at 40 ft. (12 m)	30 ft. (9 m)	20 ft. (6 m)	10 ft. (3 m)
1½-inch solid	50 (350)	250 (1,700)	4.0	2.4	0	3	7	13
			13.2	7.6	0	14	25	50
			34.0	19.9	1	28	65	96
1-inch solid	50 (350)	200 (1,400)	4.0	2.4	0	1	5	11
			13.2	7.6	0	5	21	42
			34.0	19.9	0	18	55	96
Fog (straight)	100 (700)		4.0	2.4	0	0	3	7
			13.2	7.6	0	0	13	25
			34.0	19.9	0	0	36	66
Fog (10° angle)	100 (700)		4.0	2.4	0	0	0	0
			13.2	7.6	0	0	0	0
			34.0	19.9	0	0	0	0

NOTE: Current exceeding 3 milliamps is unsafe.

1 milliamp = tingling sensation
1 - 10 milliamps = sensation of shock
5 - 25 milliamps = painful shock and muscular contraction
25 - 50 milliamps = severe muscular contraction and breathing problems
50 - 200 milliamps = ventricular fibrillation
200 milliamps = heart stops

Source: Electrical World, April 1, 1979.

the best approach is to protect exposures, making sure that the water does not reach the metal, and allow the fire to burn itself out.

If no other means is at hand, magnesium fires may be controlled by using water, if the conditions are right. The following, reprinted from NPFA Standard No. 48, *Storage, Handling and Processing of Magnesium,* will give an idea of the precautions that should be taken:

"624. Water:

"(a) Small streams of water from portable extinguishers will violently accelerate a magnesium chip fire. Water should not be used on any large chip fire since it is impossible to provide enough coolant to handle the large area of burning metal. A few burning chips can be extinguished by dropping into a bucket of water and automatic sprinklers will extinguish a typical shop fire where quantities of chips are limited.

"(b) Burning parts including castings, wrought products and fabricated structures can be cooled and extinguished with coarse streams of water applied with standard fire hoses. A straight stream scatters the fire, but coarse drops produced by a fixed nozzle operating at a distance or by use of an adjustable nozzle, flow over and cool the unburned metal. The hose streams are then worked into the fire where some temporary acceleration takes place, followed by rapid extinguishment, as water application is continued. Well-advanced fires in several hundred of pounds of magnesium solid scrap have been extinguished in less than one minute with two 1½-inch fire hoses. Water fog tends to accelerate such a fire more than cool it.

Water can be used on magnesium — with caution

"(c) *Applications of water to magnesium fires where quantities of molten metal are likely to be present should be avoided. The steam formation and possible metal-water reactions may be explosive.* [Italics in original.]

"(d) With the exercise of care to keep water from reaching molten metal, water spray applied with caution can be used safely to fight fires in combustible material in the vicinity of melting pots."

For more information concerning extinguishing agents for combustible metals, see IFSTA 101, **Forcible Entry, Rope and Portable Extinguisher Practices;** and NFPA's *Guide on Hazardous Materials.*

2

TYPES OF FIRE STREAMS & NOZZLES

NFPA STANDARD 1001
Firefighter II

4–8 Fire Streams.

4–8.1(b) The firefighter, given fire situations, for each situation shall select the proper nozzle and hose size.

4–8.2 The firefighter shall identify characteristics of given types of fire streams.

4–8.6 The firefighter shall identify four special stream nozzles and demonstrate at least two uses or applications for each.

4–8.7 The firefighter shall identify and explain foam making equipment appliances used, and shall produce a foam stream from all types of foam making appliances used by the authority having jurisdiction.

4–8.9 The firefighter, given the necessary resources, shall identify, select, and assemble those items required to develop at least three types of fire streams.

Firefighter III

5–4 Fire Streams.

5–4.1 The firefighter shall construct a diagram to identify three types of fog nozzles, and identify the major parts and trace waterflow through them.

5–4.2 The firefighter, given a selection of nozzles and tips, shall identify their type, design, operation, nozzle pressure, and flow in GPM for proper operation of each.

NFPA STANDARD 1002

3–4 Hydraulic Calculations.

3–4.1(h) The fire apparatus driver/operator shall demonstrate the principles of friction loss as they relate to use of various nozzles.

SIZES AND TYPES OF FIRE STREAMS

Fire streams are identified by size (volume of water flowing per minute) and type. Table 2.1 lists the sizes of fire streams and their use.

TABLE 2.1
Fire Stream Sizes

Volume	Use and Type
40 gpm (150 L/min)	Booster, handline Solid, fog, broken
40 - 125 gpm (150 - 470 L/min)	1½-inch handline Solid, fog, broken
125 - 350 gpm* (470 - 1,125 L/min)	2½-inch handline Solid, fog, broken
350+ gpm (1,125+ L/min)	Master streams (usually siamesed 2½-inch or larger) Solid, fog, broken

*Nozzles with flows exceeding 350 gpm (1,125 L/min) are not recommended for handlines.

Table 2.2 shows standard diameters of rubber-lined fire hose with the flows they are capable of carrying without excessive friction loss when the hose lengths do not exceed 600 feet (180 m).

The three major types of fire streams, as implied in Table 2.1, are the solid, fog and broken. Each of these types are specific patterns of water for specific jobs.

TABLE 2.2
Maximum Efficient Capacity of Rubber-Lined Hose
(Length less than 600 feet)

Hose Diameter in inches	1	1½	2½	2¾	3	3½	4	5	6
Capacity in U. S. gpm	30	100	300	400	500	750	1,000	1,700	2,500

INFLUENCING FACTORS

Modern fire fighting equipment, fire hose with improved smooth interior, calibrated nozzles and adequate water supplies are all available to produce good fire streams. Various kinds of fire streams are required for the control of structural fires, flammable liquid fires and fires in which energized electrical equipment is involved.

The coordination and use of available water, fire equipment, fire apparatus and human ability are essential for the production of effective fire streams.

Available Water

Water that is available to produce or develop fire streams may be taken from a static supply, tank trucks or a distribution system. The distribution system should be sufficient to supply all the water that is needed for domestic and fire requirements, plus a reserve for heavy demands. For further information concerning water supplies consult IFSTA 205, **Water Supplies for Fire Protection.**

Fire Equipment

The logical place to start when considering fire equipment as a factor in producing fire streams is at the source of water supply, which is usually a fire hydrant. A sufficient volume of water may not be obtained from a hydrant if the diameter of the hydrant valve opening is too small, the area of the barrel is below that recommended for the valve opening or adequate waterways are not provided.

Fire hose condition affects fire stream

The second item of equipment that influences fire streams is fire hose and its tendency to offer resistance to the flow of water. Causes of resistance such as hose size, protruding gaskets, rough lining and hose couplings, together with existing kinks, bends and elevation, have previously been discussed.

The third item of equipment to be considered as a factor in producing fire streams is the nozzle. The purpose of a nozzle is to give a fire stream forward velocity and pattern. The stream begins to take shape at the point where the nozzle is attached, and the shape of the stream is formed in the nozzle. If a nozzle is damaged its effectiveness in shaping the stream will be reduced.

Fire Apparatus

The fire department pumper should be considered as a major contributing factor for obtaining the desired waterflow and pressure. Many factors may limit a pumper's ability to perform to its maximum potential, but under favorable conditions a pumper should be capable of delivering its rated capacity.

Human Ability

Good streams depend on well-trained personnel

The best water supply, fire apparatus and fire equipment may be provided and still not produce good fire streams if personnel are poorly trained. One human error is using a supply hose or intake hose that is too small. Overtaxing a pumper with too many hoselines or with large-capacity nozzles that demand a total volume greater than the rated capacity of the pump is another human error. A very dangerous error is that of producing excessive nozzle pressure just because the apparatus has the ability to do so.

All human errors that influence fire streams are not made at the pump. Improper selection, use and application of the equipment available and failure to have adequate communication regarding the equipment being used are also human errors.

NOZZLE PRESSURE AND WATERFLOW

If a fire stream is to be of value to firefighters, it must possess the previously discussed characteristics, which are basically dependent upon nozzle pressure. The production or development of satisfactory fire streams, then, depends upon the ability of the fire pump operator to produce good fire streams by using formulas, rules of thumb and prepared tables or charts. In many cases more than one method is shown in this manual so pump operators or firefighters in general may select the method most applicable to the situation at hand.

DESIRED AND ADEQUATE NOZZLE PRESSURE

Nozzle pressure important for good streams

Tests have revealed that velocities ranging from 60 to 120 feet per second are within the practical working limits of fire streams. These limiting velocities are produced by nozzle pressures that range from 25 to 100 psi (105 to 700 kPa). A nozzle pressure of 100 psi (700 kPa) is desired and practical for all fog streams and is considered as a standard for fog nozzles. A nozzle pressure of 50 psi (350 kPa) is considered to be standard for and will produce streams from ordinary-size solid stream nozzles that have a good reach and volume. These handlines are not difficult to handle. If a greater reach and volume are desired, the nozzle pressure may be raised to 65 psi without becoming unmanageable. Above this point, solid streams become more and more difficult to handle. Nozzle pressures of 80 to 100 psi are limited to use with master stream devices. These desired pressures should be kept in mind.

NOZZLES: GENERAL DESIGN AND CONSTRUCTION

During the Freeman experiments jets from many different kinds, shapes and sizes of nozzles were compared. The results of these comparisons were surprising to the persons making the tests because they expected a polished stream of water with a clear, glassy surface and little spray. Not one of the many nozzles tested produced such a stream. After the experiments Freeman designed a standard form of tip for ordinary use. This standard tip is considered regular stock and most manufacturers use it as a basic design.

Fire Stream Nozzle Control Valves

Control valves at the nozzle enable firefighters to properly govern the distribution and effectiveness of the stream, thus regulating the amount of water. Fire stream nozzle control valves are constructed to perform a definite task. The control valves on nozzles producing fog may vary to some degree from the control valves for solid stream nozzles. The more common designs of nozzle control valves are the "ball-valve" and "rotary" types. No attempt is made here to recommend one type of shutoff valve over another, but merely to point out the types of construction.

THE BALL-VALVE CONTROL VALVE

The design and construction of the ball-valve control valve gives efficient service during fire fighting, with a minimum of effort at high pressure. The ball is suspended from both sides and seals against a neoprene seat. The ball-valve is illustrated in Figure 2.1. The seat is held captive in the nozzle body; thus, reseating the valve, if necessary, is facilitated.

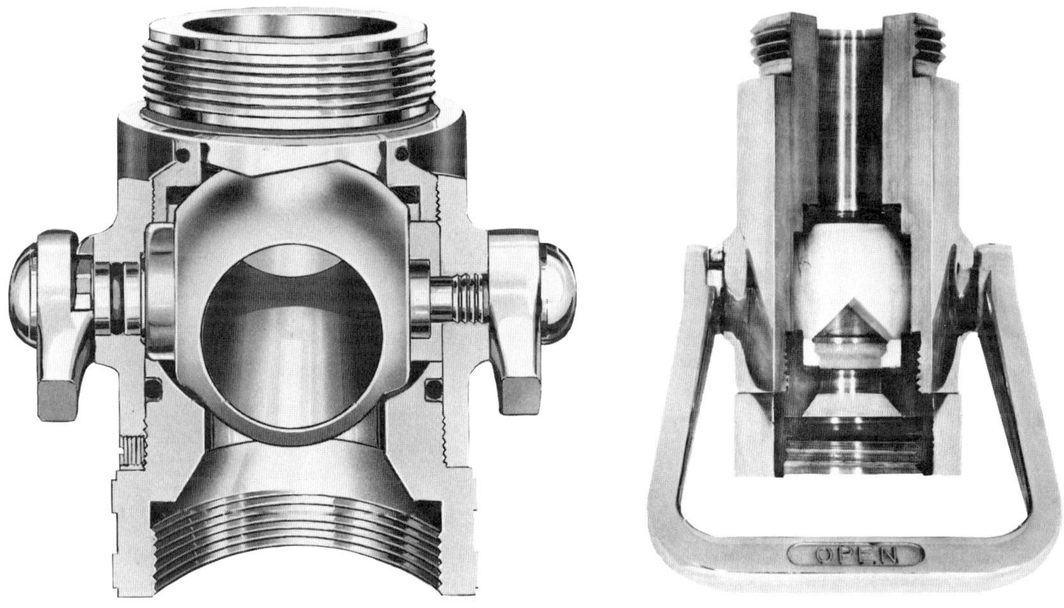

Figure 2.1 — Cutaway drawing of a ball-valve shutoff.

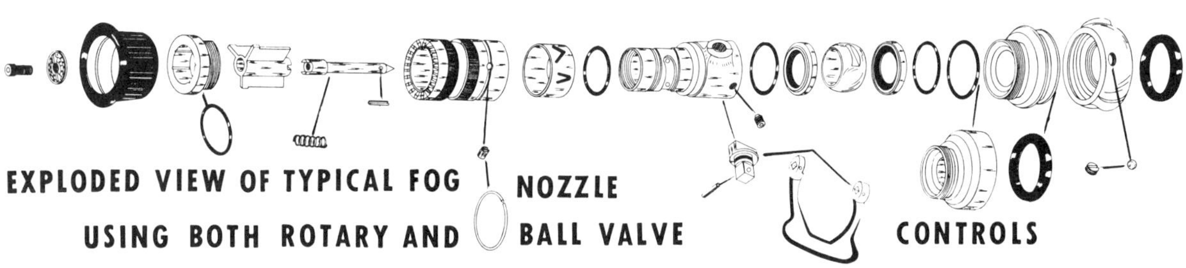

Figure 2.1a — Exploded drawing of a fog nozzle with both rotary and ball-valve controls.

THE ROTARY CONTROL VALVE

The rotary control valve used in connection with fire stream nozzles is similar to the conventional garden-hose nozzle. It consists of an exterior barrel guided by a screw that moves it forward or backward, rotating around an interior barrel.

Some rotary shutoff control valves close by means of a ground seat within the nozzle barrel. A nozzle of this type is shown in Figure 2.2. As the exterior barrel is screwed against this ground seat, the water is shut off. When this nozzle is opened, the first water to flow will be in the form of wide-pattern fog. As the exterior barrel is screwed forward, the deflection stem is drawn into the exterior barrel and produces a straight stream.

Figure 2.2 — One type of rotary control valve.

Another type of rotary shutoff control valve is shown in Figure 2.3. In this type, the deflection stem is ground to fit the ground seat at the orifice of the exterior barrel. As the exterior barrel is screwed forward against the deflection stem, the water is shut off. When this nozzle is opened the first water to flow will be in the form of a straight stream. As the exterior barrel is screwed backward, the deflection stem is pushed out of the exterior barrel and produces a wide-pattern fog stream.

Figure 2.3 — Another type of rotary control valve.

Stream Straighteners (Vanes)

The rotary motion and twisting currents at the nozzle have received considerable study and discussion. Several devices have been introduced to prevent this twisting motion so a more perfect stream is produced. These devices consist of some form of straight internal blades inserted parallel to the axis of the playpipe. These internal blades are referred to as *vanes* or *stream straighteners.* Some types of stream straighteners are illustrated in Figure 2.4

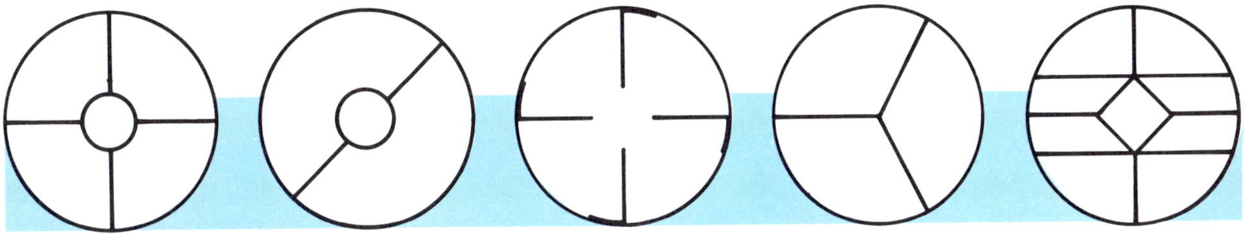

Figure 2.4 — Examples of vanes, or stream straighteners, used in fire hose nozzles to diminish water turbulence.

Careful experiments with a master stream nozzle whose stream straighteners could be removed proved that removal of these vanes had a marked negative effect upon the jet in decreasing its reach and conformity.

With the ordinary siamese, by which two, three or even four hoselines are united into one at the base of the master stream device, some arrangement is usually necessary to reduce the whirling action set up by the converging streams.

MECHANICAL PRINCIPLES OF SOLID STREAM NOZZLES

At the present time, solid stream nozzles are designed so the shape of the stream after it leaves the hose is gradually reduced in the nozzle until a point a short distance from the outlet is reached. At this point the nozzle becomes a cylindrical bore, the length of which is from 1 to 1.5 times the diameter of the bore. The object of this short, truly cylindrical bore is to round out and give the stream its shape before discharge. A smooth-finished waterway contributes to both the shape and reach of the stream. It is surprising what small differences or slight deviations the shape of the nozzle make when the tip is not choked, equipped with rings or built with other than a truly cylindrical form at the tip. A basic design used by most manufacturers of solid stream nozzles is illustrated in Figure 2.5. Figure 2.6 is an exploded drawing of one type of solid stream hand-held appliance.

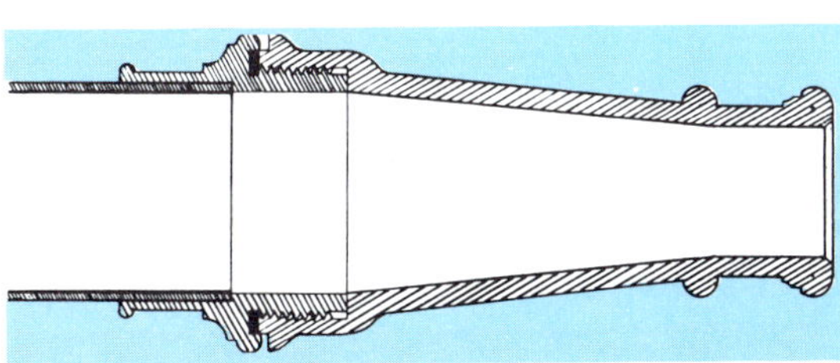

Figure 2.5 - Basic design of solid stream nozzles.

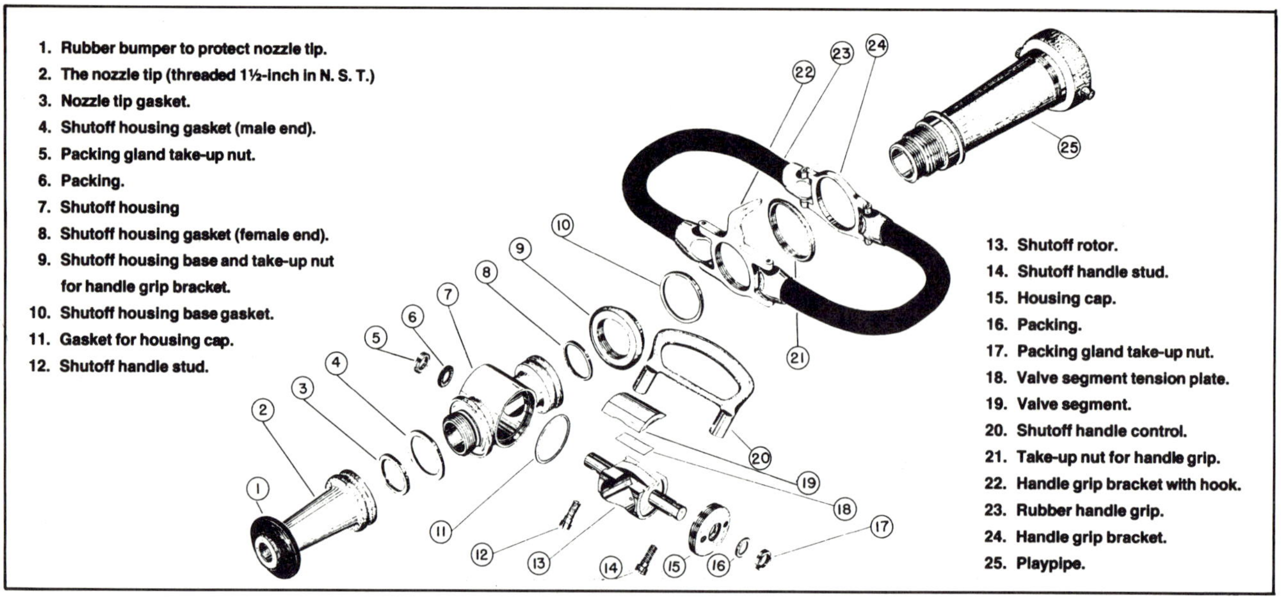

Figure 2.6 — Exploded view of solid stream nozzle assembly.

The solid stream nozzle is designed to produce a stream as compact as possible with little shower or spray. Its ability to reach areas that might not be reached by other nozzles has established the solid stream as necessary for the fire service.

Many designs and shapes of nozzles that produce solid streams are available to the fire service, and manufacturers are constantly striving for better designs. Although these nozzles differ in shape and size, they all conform to the required physical characteristics for solid stream nozzles. The shutoff valve connection to the playpipe and the shutoff valve connection to the tip should be threaded with 1½-inch (38 mm) National Hose thread. This standardization has been recommended by the National Fire Protection Association since it offers the greatest degree of flexibility in the practical problems of actual fire fighting. Based upon the Freeman experiments, streams from well-designed nozzles have effective reach as shown in Table 2.3.

Nozzle condition affects solid streams

The extreme limit at which a solid stream of water can be classified as a good stream cannot be sharply defined and is, to a considerable extent, a matter of judgment. It is difficult to say within five feet, or in some instances even within ten feet, just exactly where the stream ceases to be good. Any observations or recordings that may be made to classify solid streams reflect the characteristics of the condition of the nozzle. Observations and tests covering the effective range of fire streams have classed as good a stream that at the point of breakover:

- Has not lost continuity by breaking into showers of spray.
- Appears to shoot nine-tenths of the whole volume of water inside a circle 15 inches (0.4 m) in diameter and three-quarters of it inside a 10-inch (0.25 m) circle, as nearly as can be judged by the eye.

TABLE 2.3
Effective Range of Solid Fire Streams
(in feet)

	Nozzle Diameter (inches)									
	1		1⅛		1¼		1⅜		1½	
Nozzle Pressure	Vertical Distance[a]	Horizontal Distance[b]	Vertical Distance	Horizontal Distance	Vertical Distance	Horizontal Distance	Vertical Distance	Horizontal Distance	Vertical Distance	Horizontal Distance
20	35	37	36	38	36	39	36	40	37	42
25	43	42	44	44	45	46	45	47	46	49
30	51	47	52	50	52	52	53	54	54	56
35	58	51	59	54	59	58	60	59	62	62
40	64	55	65	59	65	62	66	64	69	66
45	69	58	70	63	70	66	72	68	74	71
50	73	61	75	66	75	69	77	72	79	75
55	76	64	79	69	80	72	81	75	83	78
60	79	67	83	72	84	75	85	77	87	80
65	82	70	86	75	87	78	88	79	90	82
70	85	72	88	77	90	80	91	82	92	84
75	87	74	90	79	92	82	93	84	94	86
80	89	76	92	81	94	84	95	86	96	88
85	91	78	94	83	96	87	97	88	98	90
90	92	80	96	85	98	89	99	90	100	91

NOTE: Ranges shown are with moderate wind (to 18 mph). Reach is greatly reduced with stronger winds.

[a]Nozzle elevated 60 to 75 degrees.
[b]Nozzle elevated 30 to 35 degrees.

Source: John R. Freeman, *Transactions*, American Society of Civil Engineers, Volume 21.

- Would probably be stiff enough to attain in fair condition the height required even though a fresh breeze is blowing.

Water from a nozzle producing a jet to conform with these characteristics would be thrown much farther than the breaking point, but beyond this point it tends to be in the form of a broken stream easily carried away. These characteristics and the point of breakover are illustrated in Figure 2.7.

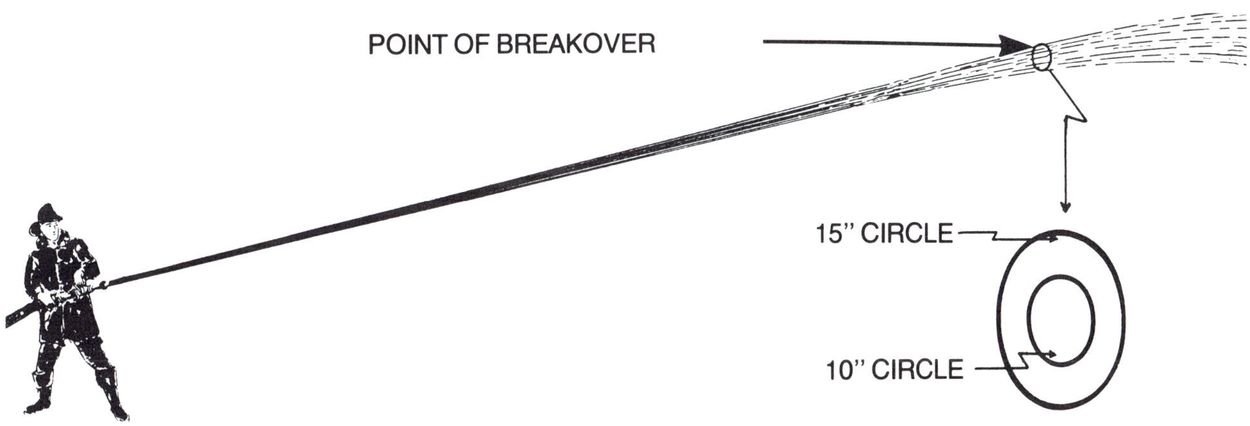

Figure 2.7 — Point of breakover

Flow Capacity of Solid Streams

The rate of discharge of a fire stream is measured in gallons per minute (gpm) or liters per minute (L/min). Flow from a nozzle is determined by the velocity of the stream and the size of the discharge opening. These two factors are influenced by other forces, and for this reason they will be considered separately.

Velocity is the speed of an object in a given direction. The speed of a stream of water is developed by pressure upon the water. Since water is practically incompressible, the water will flow at a certain rate of speed.

The gallons-per-minute flow-capacity of a nozzle increases by four times when the diameter of the discharge opening is doubled. Although it might appear that the capacity of two 1-inch (25 mm) nozzles is equivalent to the capacity of a 2-inch (50 mm) nozzle this is not true. As previously illustrated in Figure 1.30, four 1-inch nozzles are required to allow the same capacity under the same conditions as a 2-inch nozzle. On the other hand, if two 1-inch nozzles are being used and if a larger nozzle of the same capacity is desired a 1⅜-inch (35 mm) nozzle would be the proper size. A nozzle of a certain size will discharge just so much water at a definite stream velocity. To increase the discharge a firefighter must either use a larger nozzle or increase the stream velocity. One basic rule concerning flow capacity of a solid fire stream states: *The nozzle tip size should not be larger than one-half the diameter of the hose.* For example, the maximum size of the nozzle tip for 2½-inch (65 mm) hose is 1¼ inches (32 mm).

Hose/nozzle tip ratio for solid streams

Effective Reach of a Solid Stream

Just as a solid fire stream will deliver a definite discharge under given conditions, it will also travel a definite distance from the nozzle. The reach of a solid fire stream is the distance that a stream can be effectively thrown from a nozzle. Two forces, gravity and friction of the air, tend to decrease this reach. These opposing forces must be overcome by the forward velocity and the volume of the stream. Since the stream is beyond mechanical control after it leaves the nozzle, its speed while passing through space and the distance it will travel are determined by these two factors.

If the stream is encased within a pipe and put under 50 psi (350 kPa) pressure at the base of the pipe, the water will fill to a height of 115 feet (35 m). In other words, if one psi will raise water in the pipe 2.3 feet (70 cm), then 50 psi will raise the water 115 feet. If the same stream of water is not encased in a pipe but is directed vertically into the air the performance changes. After leaving the tip the stream tends to hold its shape for a considerable distance and then begins to flare out and break into drops. These drops of water soon lose their momentum and fall to the

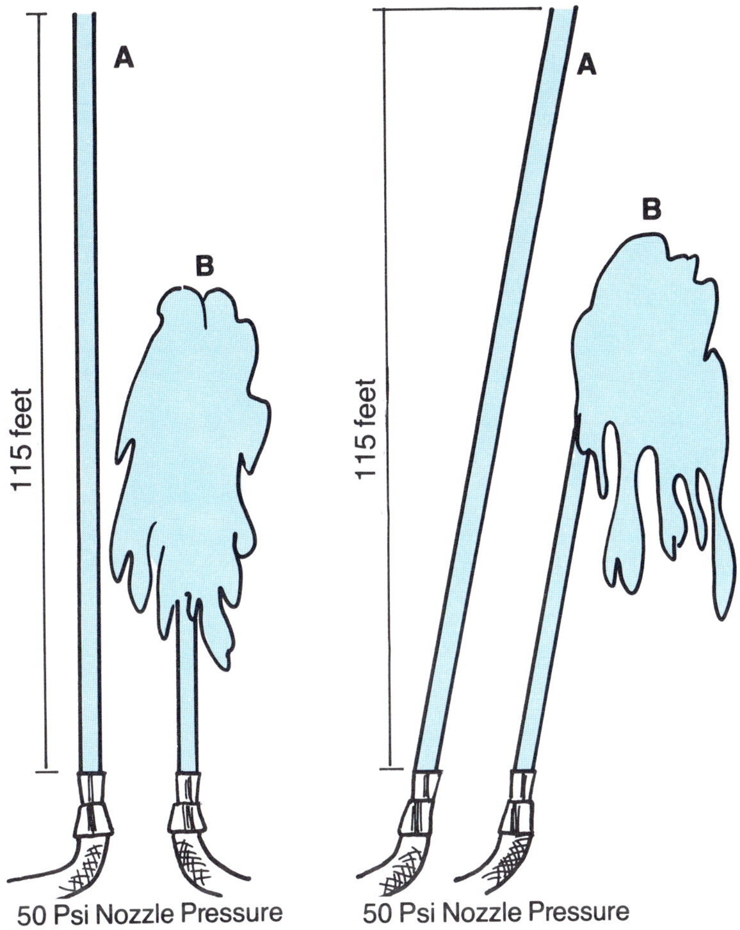

Figure 2.8 — Drawings A represent water encased in a pipe. Drawings B represent free water and the effects of gravity and air friction. *Source:* NFPA, *Fire Protection Handbook,* 14th ed.

ground. This breaking away continues until only the inner part of the column reaches the maximum height. The friction of the air on the outer surface of the water column breaks the stream, and the force of gravity pulls the drops to the ground.

If the pipe that encases the column of water is inclined and if its length is increased to the same vertical height, the same water column of 115 feet may be obtained. This equal water level may be explained by the sixth principle of pressure on fluids: *The pressure of a liquid on the bottom of a vessel is independent of the shape of the vessel.* By removing the pipe and inclining the nozzle, as shown in Figure 2.8, the water falls short of the desired height because of the air that has been introduced as an opposing force. Tests with solid streams reveal the following characteristics of reach:

- The greater the stream velocity at the nozzle, the greater the reach of the stream.
- The greater the volume of water with the same stream velocity, the greater the reach of the stream.
- The closer the directed stream reaches an angle of 32 degrees to the earth's surface, the greater the reach of the stream.

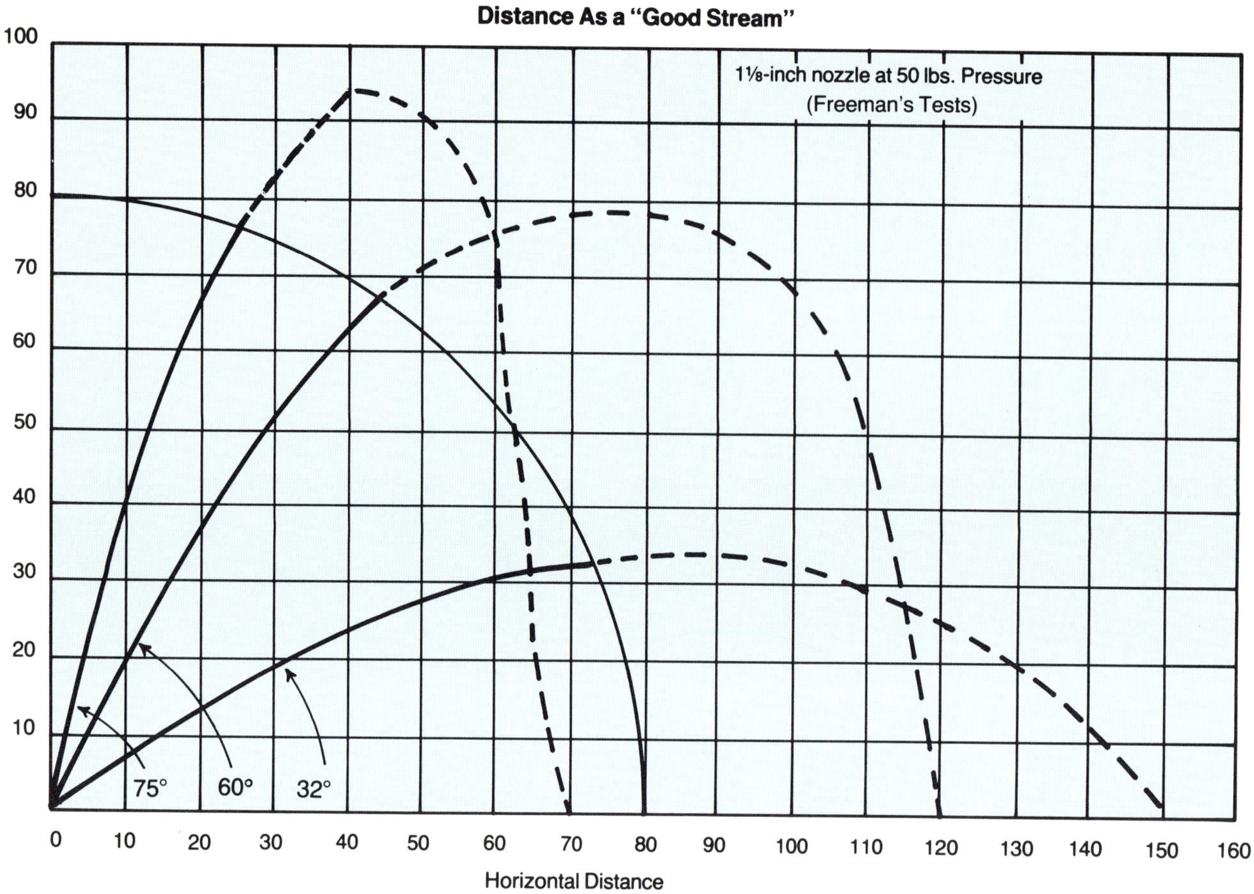

Figure 2.9

Tables that have been published on ranges of fire streams are usually computed for streams while the air is comparatively still. If a breeze is blowing, the reach is materially reduced and, for that reason, it is desirable to have some margin of safety. Comparative horizontal distances for effective solid streams in still air are shown in Figure 2.9.

FOG STREAMS

The terms *fog nozzle* and *fog stream* are commonly used in the fire service as acceptable descriptive terms of a patterned stream composed of droplets, and will be used in this manual.* The adjustable fog nozzle is the most popular method of making a fog stream. These nozzles may be adjusted from a wide-angle fog pattern to a straight stream pattern. It should be understood that a *straight stream* is a pattern of the adjustable fog nozzle whereas a *solid stream* is discharged from a smooth-orifice nozzle.

---

*Here is where the difference between a fog stream and a broken stream lies: the fog stream has a definite pattern, whereas the broken stream does not always have a definite pattern and is usually composed of larger drops.

Mechanical Principles and Design of Fog Nozzles

Any jet of water that has a given velocity and is not mechanically retarded in its path will remain in a solid mass and not lose continuity until it strikes an object, is overcome by gravity or is changed by friction with the air. For a stream of water to be broken into finely divided particles it must be bombarded with sufficient velocity against an obstruction to shatter the mass. The angle at which the stream of water is deflected from an obstruction determines the reduction in forward velocity of the stream. Therefore, a wide-angle deflection produces short-reaching fog, and a narrow-angle deflection produces longer-reaching fog. A fog stream may be produced by deflection at the periphery, by impinging jets of water or by a combination of these.

DEFINITIONS

Periphery — The line bounding a rounded surface; the outward bounds of a thing distinguished from its internal regions.

Deflection — A turning or state of being turned; a turning from a straight line or given course; a bending; a deviation.

Impinge — To strike or dash about or against; clashing with a sharp collision; to come together with force.

If water can be made to travel at angles to the direct line of discharge the reaction force may be made to more or less counterbalance each other and reduce the nozzle reaction. This balancing of forces is the reason why fog streams can be handled more easily than can solid streams.

The desired performances of fog stream nozzles are dependent upon

- The quality of the fog stream.
- The reach of the fog stream.
- The space occupied by the fog stream.
- The ease of handling the fog stream.

These performances are usually judged by

- The amount of heat that a fog stream will absorb.
- The rate at which the water is converted to steam.

These last two factors should be considered when one fog stream nozzle is compared with another because it is from these two performances that a decrease in temperature, the displacement of smoke and gas, and a minimum of water damage are obtained.

Factors that should be considered when selecting a fog nozzle are the type, the size and weight of the nozzle and the simplicity of its operation. The design of the fog nozzle controls certain factors that influence the fog stream. A fog nozzle may be considered to have a perfect design that meets all requirements, yet without the required pressure at the nozzle it may prove to be inadequate. Low nozzle pressure will adversely affect the quality of the stream.

One question that has been of great concern to the fire service is the matter of operating pressures. When only the design of the nozzle and the pressure at the nozzle are considered it is easy to overemphasize operating pressures. As pressures are increased, factors appear that influence the development of the fog stream pattern. The velocity of the jet, the size of the water particles and the volume of water discharged are more important than any excess in pressure that may be necessary to produce a stream that will reach a desired objective.

The Velocity of the Stream

The exact measurement of the speed of the water toward its objective is of little value to firefighters. It does not matter whether it takes one or five seconds for a particle of water to travel from the nozzle to the fire, as long as the stream reaches the objective in the desired pattern. When a stream is driven against an obstacle the forward velocity is partially reduced in proportion to the shape of the obstacle and the angle at which the stream hits. If the surface of the obstacle is at a right angle to the stream the entire forward velocity will be lost, but if it is less than a right angle the loss will be in proportion to the angle. As soon as water is discharged from a nozzle it is carried forward by its momentum and downward by the force of gravity, and its motion is retarded by friction with the air. Obviously, the faster the water is traveling forward the farther it will travel before it is pulled to earth by gravity. Forward velocity, therefore, is one factor that governs the reach of a fog stream.

Velocity less important than reach

The effect that nozzle design or adjustment has upon a fog stream may be illustrated by operating two identical adjustable nozzles at identical pressures. Since the pressures are equal at both nozzles it can be assumed that water is being discharged from both at the same velocity, but the difference in the adjustment of these nozzles demonstrates a marked difference in the two streams. The fog stream in Figure 2.10 appears to have considerably less reach and forward velocity than the one in

Figure 2.11. The design of the nozzle permits adjustment for different stream patterns. This adjustment changes the angle of the discharge orifice; a smaller angle increases the forward velocity of the stream. Fog streams should be evenly divided into a fine spray with a uniform discharge pattern around the "cone" regardless of the nozzle setting.

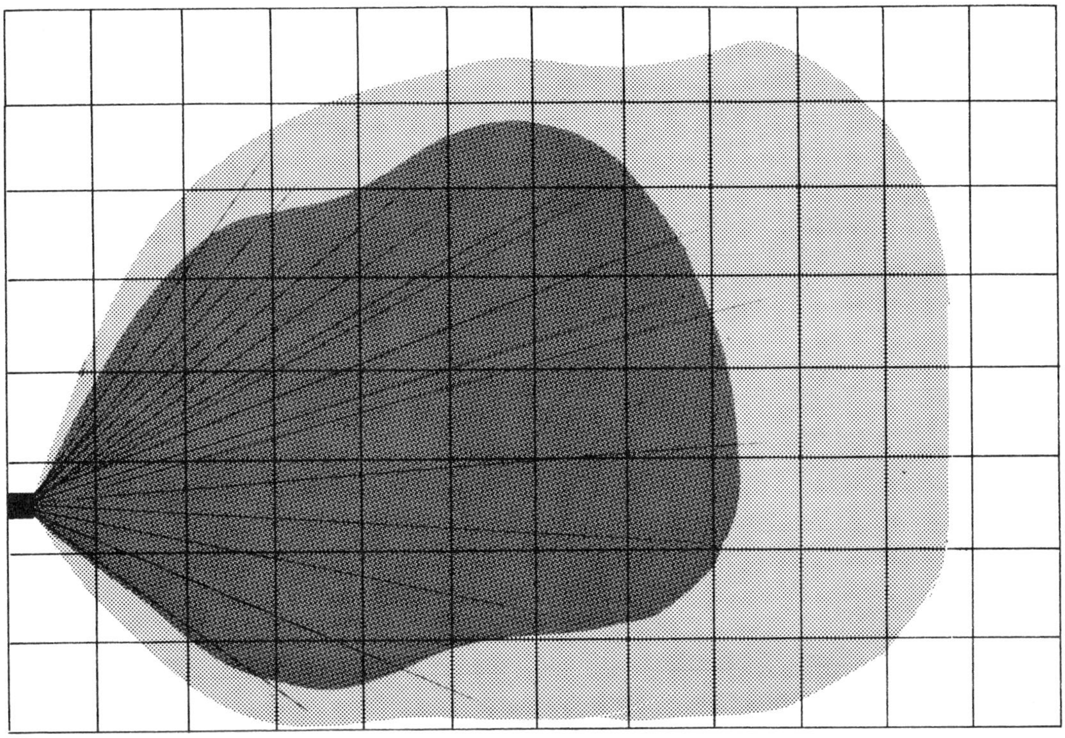

Figure 2.10 — Wide-angle fog has less reach and forward velocity.

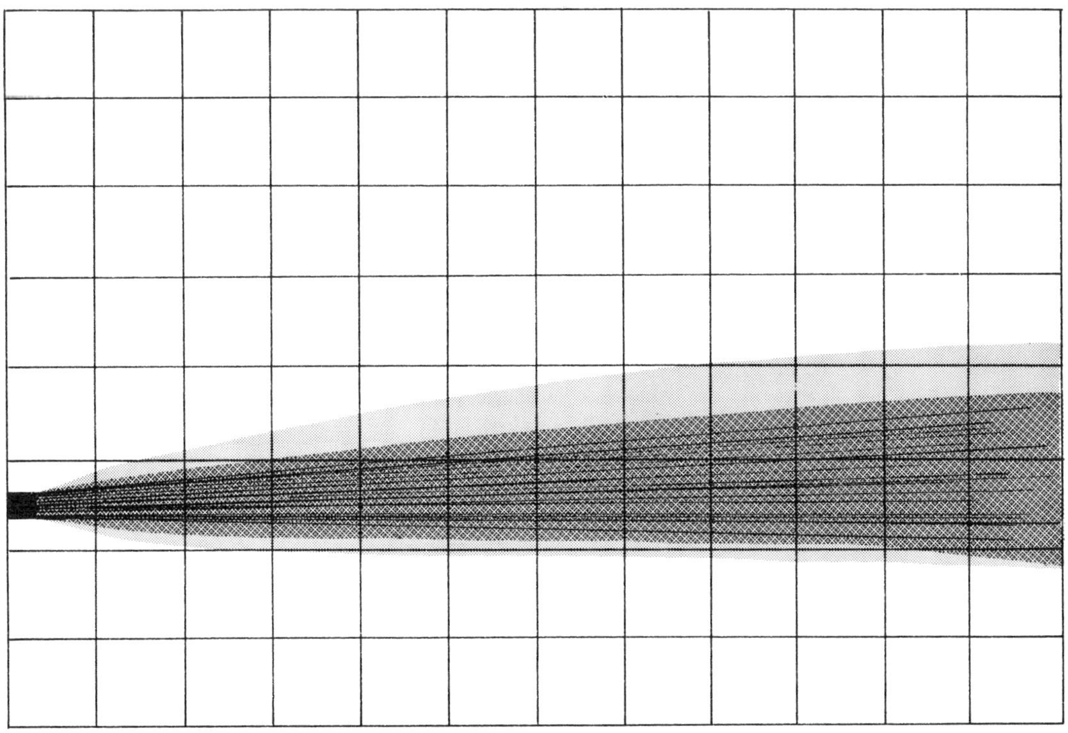

Figure 2.11 — Narrow-angle fog has greater reach and forward velocity.

A short-reaching fog has little forward velocity. A longer-reaching fog has considerable forward velocity and its reach will vary in proportion to the pressure applied. Unsatisfactory longer-reaching fog discharge patterns are often caused by insufficient nozzle pressures. There is a maximum reach, however, to any longer-reaching fog pattern, and once the nozzle pressure is sufficient to produce a satisfactory reach further increases in nozzle pressure will have relatively little effect upon the stream except to increase the volume.

The Size of the Water Particles

Air friction affects fog streams

Any stream of water is retarded in its path through the air because of the friction between the outer surface of the stream and the air. A fog stream is composed of droplets of water surrounded by air. The air is entrained in the stream and becomes part of the stream. A fog stream is retarded by the friction between the outer surface of the fog cone and the surrounding air, turbulence within the fog stream, and the process of entraining air. If two fog streams are of the same diameter and the water droplets have the same initial velocity, the one containing the most water will have the greatest reach because it will have a greater momentum. The size of the droplets will have some effect on the reach of the fog stream because of the amount of entrained air. Since fog streams have a larger diameter in comparison with solid streams, there is more area subject to friction with the air and its forward velocity is retarded more rapidly.

Water may be divided so finely that the air appears to be saturated with a fine mist and the small water particles appear to be suspended in the air (Figure 2.12). Water particles in this form are so fine that they may be carried away from the fire by air currents and never penetrate the fire area. The result is that they do not have an opportunity to absorb heat with sufficient rapidity and quantity to be effective. To maintain momentum and obtain desired reach and penetration a spray of water as shown in Figure 2.13 may be desirable and effective. Fog patterns should be sufficiently heavy to work into a moderate wind without being destroyed.

The Volume of Water Discharged

Another factor determined by the design of the nozzle and the pressure at the nozzle is the volume of water discharged. Fog streams are superior to solid streams when heat absorption qualities are compared, but fog streams must have sufficient volume to penetrate the heated area. Water cannot absorb its full capacity of heat unless it is completely converted into steam. If a low-volume nozzle producing finely divided particles is used where the heat is generated faster than it is absorbed the fire might be checked to a

Figure 2.12 — Fine water particles such as these are easily affected by wind and air currents.

Figure 2.13 — Larger water particles have greater resistance to the effects of air currents; thus, greater penetration and reach.

certain extent, but extinguishment probably would not be accomplished. A fog stream must deliver a volume of water that is sufficient to absorb heat more rapidly than the heat is generated.

The Reach of the Fog Stream

The reach of a fog stream is necessarily less than that of a comparable solid stream. This is because of the way fog streams are formed. Impinging or deflecting a stream of water lessens its velocity, and the droplets formed are affected more by friction with the air. Nonetheless, the fog stream is an important fire fighting tool for enclosed fires. What the fog stream lacks in reach is made up for by the larger space it occupies and the greater amount of water surface exposed to the heat.

The Space Occupied by the Fog Stream

The size of the area occupied by the fire within a building determines the required reach of the fog stream. If a fire condition requires a far-reaching fog stream to cover the entire heated area, the space occupied by the fog stream will differ greatly from a fog stream pattern with less reach. The maximum heat absorption from the space occupied by a fog stream can be most clearly shown by illustrations. A potentially far-reaching fog stream pattern entering into an area of approximately 15 feet square (21 m^2) is shown in Figure 2.14. This type of fog stream pattern under these conditions does not occupy much space. Furthermore, this fog stream cannot absorb heat with maximum efficiency, because a large percentage of the water is being projected against the ceiling and opposite wall where its heat-absorbing ability cannot be used. If the same nozzle is set to occupy a larger space for this particular area very little water runs down the opposite wall and a larger percentage of the water fog will be converted into steam (Figure 2.15).

Figure 2.14 — This fog stream does not occupy very much space, so it cannot absorb much heat.

Figure 2.15 — This fog stream occupies more space and can absorb more heat than the stream in Figure 2.14.

If fire conditions are such that the size of the fire area is considerably larger, however, one or more far-reaching fog streams may be needed to project sufficient water into the heated area for maximum heat absorption. A fog stream pattern with a wide space coverage will reduce the temperature in the immediate area, and may absorb little heat beyond its reach, but the displacement of the heated gases by the stream may be of assistance.

TYPES OF FOG NOZZLES

The Periphery-Deflected Stream Nozzle

One of the two basic designs of fog nozzles is the periphery-deflected stream nozzle. Water flowing through this type of nozzle is deflected from the periphery of an inside, circular stem (Figure 2.16). This deflected water is deflected again by the exterior barrel. The spatial relation of the deflecting stem and the exterior barrel determines the shape of the fog stream. The patterns of most of these nozzles are adjustable, although some are preset. The adjustment is made by turning the exterior barrel. The principle is the same as that of the common garden-hose nozzle.

There are refinements in design with some periphery-deflected nozzles. One is the incorporation of toothlike protrusions around the inside rim of the exterior barrel. These deflect some of the water to the center of the water cone. Another refinement is to use impinging jets that produce a cone of water within the outer cone and thus produce a more homogenous stream.

Figure 2.16 — The relative positions of the deflection stem and the exterior barrel of a rotary-controlled fog nozzle determine the stream pattern.

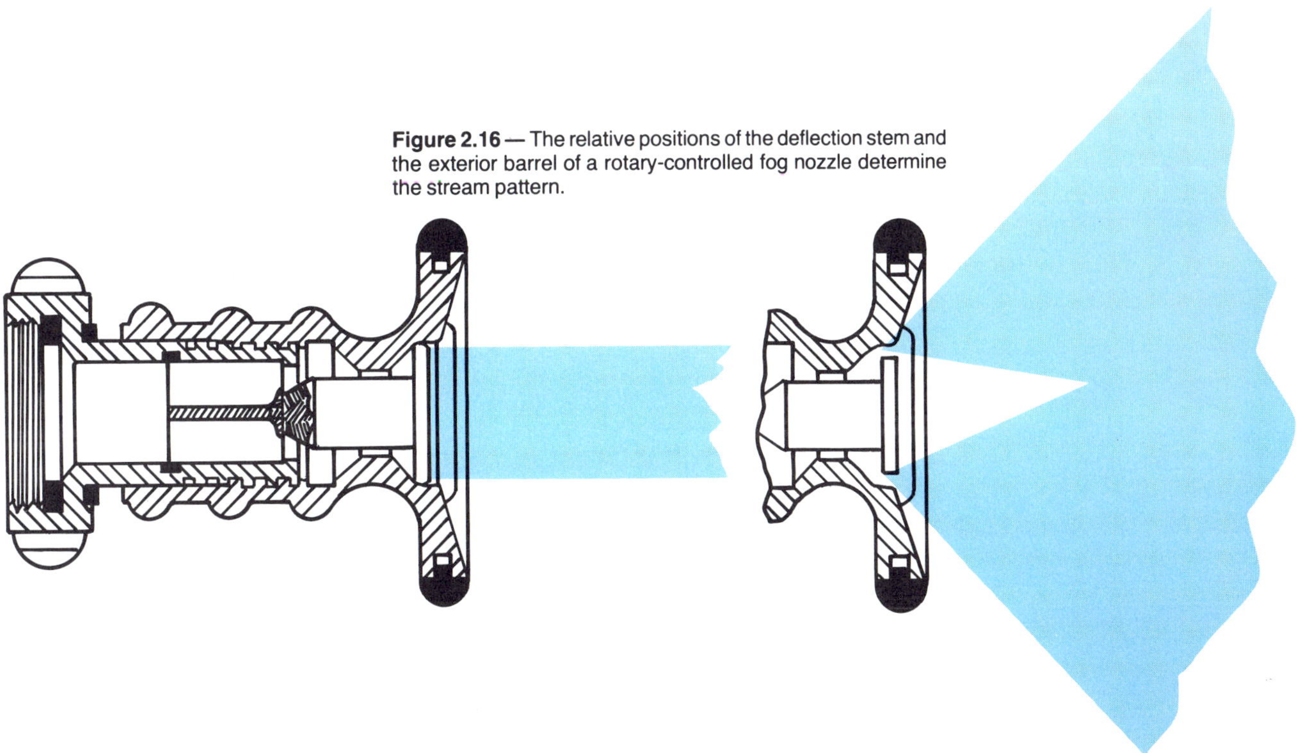

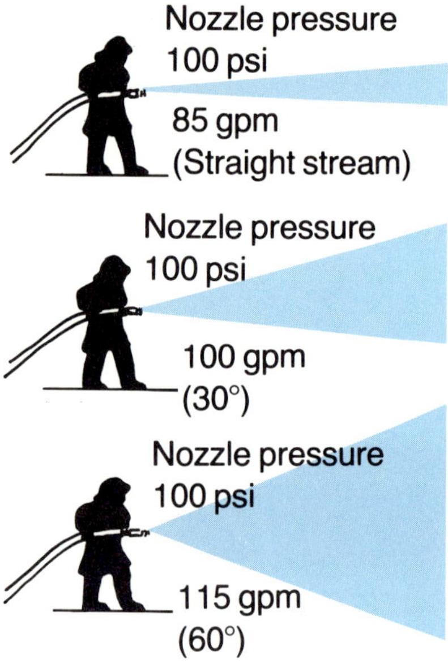

Figure 2.17 — The volume delivery of a nonconstant-gallonage nozzle changes with the pattern setting.

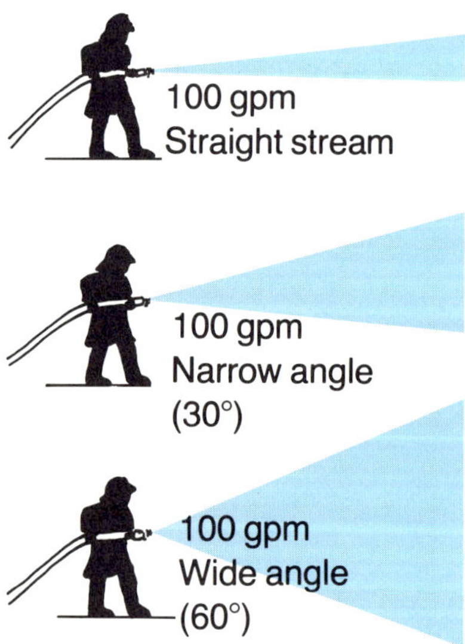

Figure 2.18 — A constant-gallonage nozzle delivers a constant flow whatever the pattern setting.

Nonconstant-Gallonage Nozzles

The pattern of a *nonconstant-gallonage* nozzle (also called variable-gallonage) is adjustable, and any change in pattern causes a change in gallonage. These nozzles are flow-rated at a narrow setting (about 30°) at 100 psi base-of-the-nozzle pressure. If the nozzle pressure is held at 100 psi and the pattern is changed to a straight stream, the gallonage will be reduced approximately 15 percent. If the pattern is changed from 30° to the widest possible pattern, the gallonage will increase approximately 15 percent.

Constant-Gallonage Nozzles

The *constant-gallonage* nozzle (also referred to as constant-flow) discharges the same volume per minute whatever the pattern setting. These nozzles usually have three settings: narrow-angle, wide-angle and straight. As the exterior barrel is rotated to change the pattern, the space between the deflecting stem and the internal throat remains the same, so the same quantity of water will pass through.

Adjustable-Gallonage Nozzles

The *adjustable-gallonage* nozzle is a constant-flow nozzle with a refinement: the firefighter at the nozzle can choose from a number of flow settings. This feature allows the firefighter to select the gallonage to suit existing conditions. For instance, if the nozzle pressure is affected for some reason, the nozzleman can adjust the nozzle to compensate. At the same time, the nozzleman should know what flow rate will be personally protective and provide ample gallonage for fire extinguishment.

Automatic Nozzles

Automatic nozzles (also referred to as constant-pressure or pressure-regulating) are basically constant-gallonage with pattern change, but not with pressure change. Their primary effect is to maintain an essentially constant nozzle pressure. If for some reason the gallonage to the nozzle changes, the automatic nozzle will maintain the nozzle pressure and the pattern. This feature is made possible by a baffle that moves automatically, varying the spacing between the baffle and the throat. A stream from an automatic nozzle can look good, but there might not be sufficient water for extinguishment or protection. For this reason, departments that use automatic nozzles should determine a minimum operating pressure as part of standard operating procedures when an automatic nozzle is in operation.

In the calculations for manual fog nozzles (100 psi operating pressure) presented in this text, calculated pump pressure will almost always fall between about 125 and 250 psi, usually 150 and 200 psi. Thus, a quick standard procedure for automatic

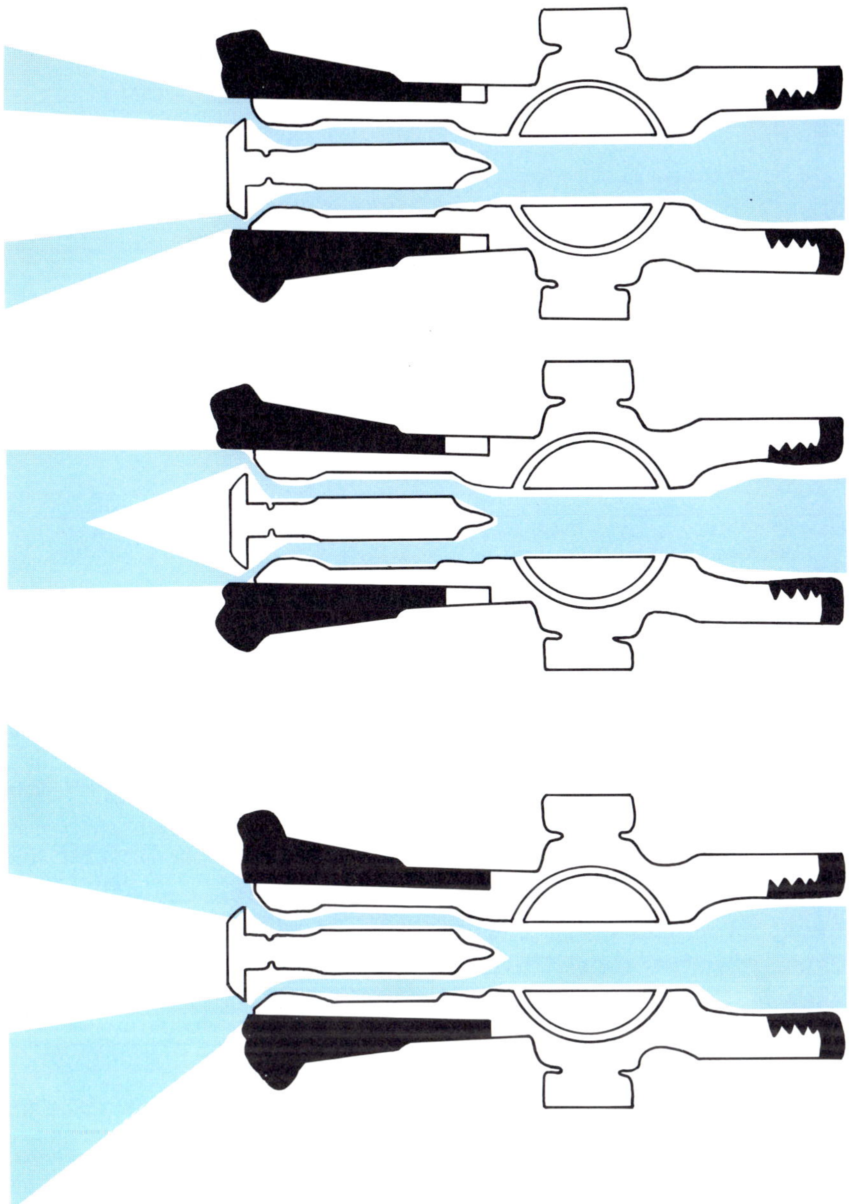

Figure 2.19 — Cutaway views showing relative positions of deflection stem and exterior barrel of a constant-gallonage nozzle at three stream patterns.

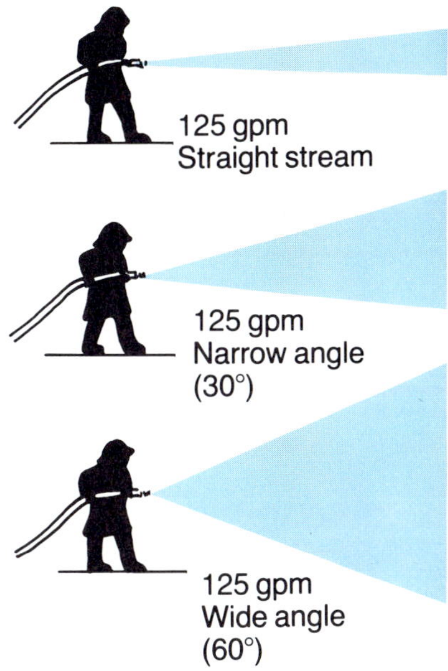

Figure 2.20 Adjustable-gallonage nozzle.

nozzles becomes apparent: Start with a pump pressure (predetermined according to most frequent layouts and available manpower) within the above range. As time and other variables permit, adjust the pump pressure according to needs, availability and calculations. The predetermined pump pressure must establish a minimum below which an inadequate flow will be delivered.

Should supply be inadequate the maximum volume is that which can be achieved without the pump "running away," in which case the throttle is backed off slightly from the point where the pump starts to speed up or become erratic or where supply hose becomes soft and begins to collapse. This will provide the maximum flow possible for the particular situation until the water supply can be augmented.

Figure 2.21 — Adjustable-gallonage nozzle. Note flow settings from which nozzleman may choose. The chosen flow will remain constant whatever the pattern setting.

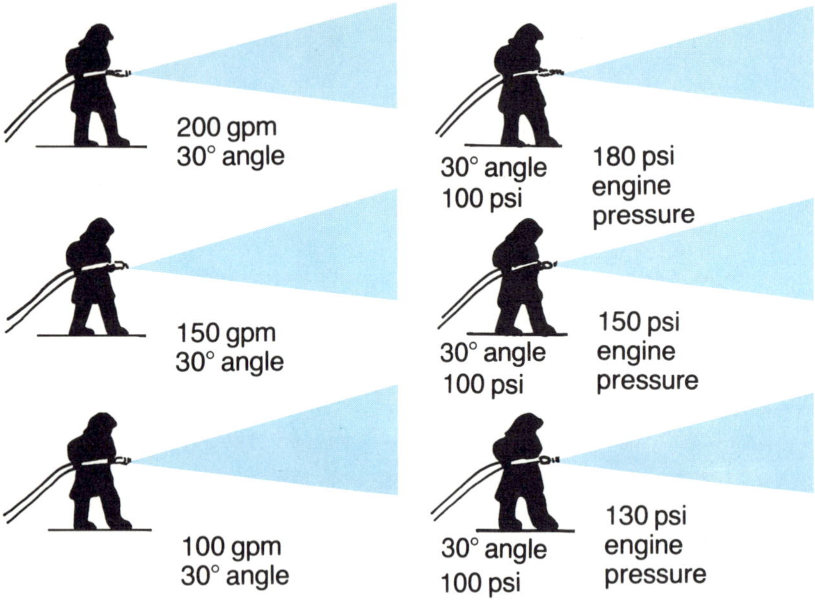

Figure 2.22

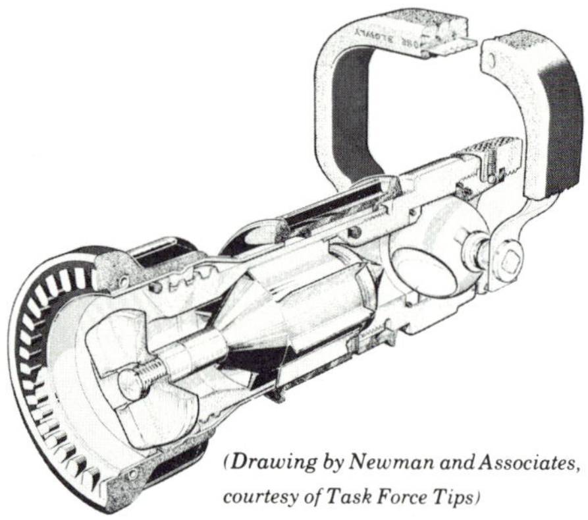
(Drawing by Newman and Associates, courtesy of Task Force Tips)

Figure 2.22a — Cross-sectional drawing of an automatic nozzle.

An automatic nozzle will serve as a pressure regulator within its flow limits for the pumper as lines are added or shut down; thus, all available water may be used continuously, if desired.

The Impinging Stream Nozzle

An impinging stream nozzle drives several jets of water together at a set angle for the purpose of breaking the water into finely divided particles. The angle setting and design of the nozzle usually produce a short-reaching fog pattern, but a long-reaching pattern is possible. A longitudinal section of one type of impinging deflection stream nozzle is shown in Figure 2.23.

BROKEN STREAMS

A broken fire stream is a stream of water that has been broken into coarsely divided drops. A solid stream may become a broken stream because of reacting forces, and it may be desirable for a broken stream to break near the fire to sprinkle the burning material. Other means, such as rotary distributor nozzles, flat

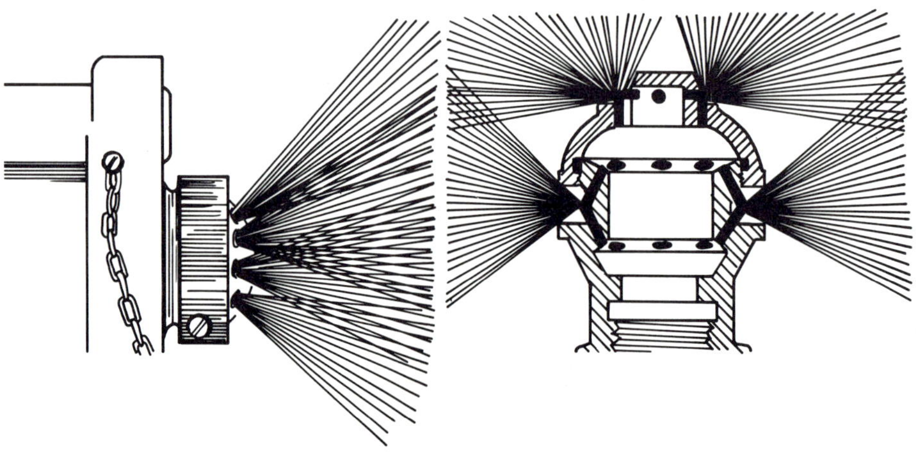

Figure 2.23 — Two types of impinging stream fog nozzles.

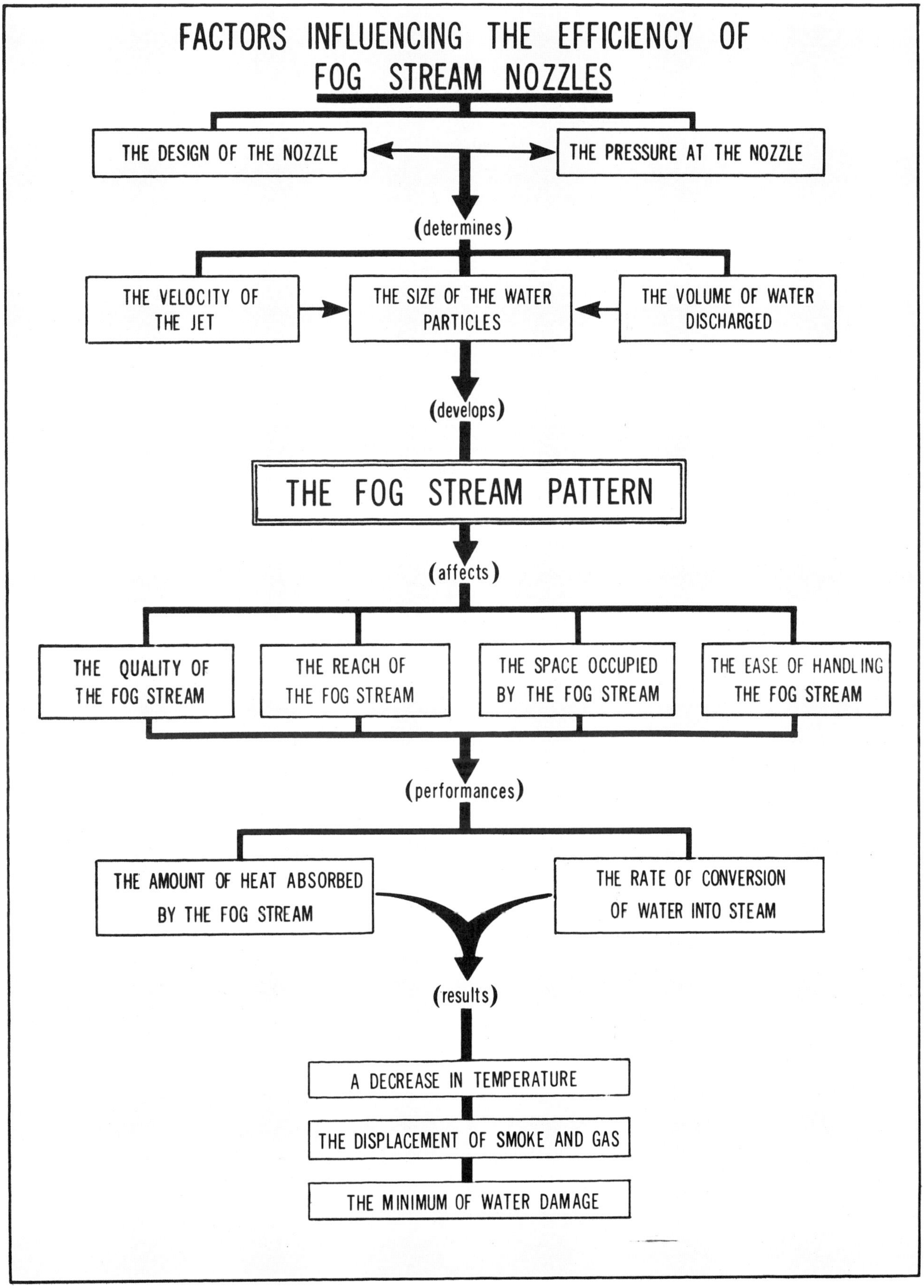
FACTORS INFLUENCING THE EFFICIENCY OF FOG STREAM NOZZLES
THE DESIGN OF THE NOZZLE
THE PRESSURE AT THE NOZZLE
(determines)
THE VELOCITY OF THE JET
THE SIZE OF THE WATER PARTICLES
THE VOLUME OF WATER DISCHARGED
(develops)
THE FOG STREAM PATTERN
(affects)
THE QUALITY OF THE FOG STREAM
THE REACH OF THE FOG STREAM
THE SPACE OCCUPIED BY THE FOG STREAM
THE EASE OF HANDLING THE FOG STREAM
(performances)
THE AMOUNT OF HEAT ABSORBED BY THE FOG STREAM
THE RATE OF CONVERSION OF WATER INTO STEAM
(results)
A DECREASE IN TEMPERATURE
THE DISPLACEMENT OF SMOKE AND GAS
THE MINIMUM OF WATER DAMAGE

Figure 2.24

spoon-bill nozzles, or directing two solid streams together in midair, may be used to produce broken streams (see "Broken Streams," Chapter 4). Regardless of the method used to produce a broken stream such streams have a definite purpose in fire fighting. The droplets are larger than those of a fog stream and have great penetration, so the broken stream can be useful where neither a fog stream nor a solid stream would be best. A broken stream used as a water curtain is effective exposure protection. Figure 2.25 shows several broken-stream nozzles.

A broken stream should cover the most space possible and yet afford enough heat absorption to extinguish fire. These broken streams are useful when fighting a basement fire through the floor above or when fighting a fire that is in a partition. The volume of a broken stream should be comparable to that of a solid stream. Where broken streams are used water damage is not generally given as much concern as extinguishing the fire, since a considerable amount of water is projected over the general area.

Water curtains protect exposures and personnel

A water curtain is a fire stream that takes, as nearly as possible, the shape of a curtain or fan between a fire and combustible material. A water curtain is sometimes desirable to protect firefighters from heat. One who has had the experience of following a fog nozzle into a heated room or against a fire can readily appreciate this use. The fundamental requirement of a water curtain is to protect or shield an exposure. The curtain must cover a wide area and be reasonably heavy to fulfill this requirement. (Tests have indicated that a water curtain between the fire and combustible material is not as effective as the same amount of water flowing over the surface of the combustible material. See "Exposures," Chapter 4.)

MASTER STREAMS

The term *master stream* is applied to any fire stream that is too large to be controlled without mechanical aid. Safety precautions, such as tying off, should be employed. A master stream may be either solid or fog and is produced by using special nozzles, paralleled hoselines and large-capacity pumps. A master solid stream requires a large, smooth nozzle; a master fog stream requires a fog nozzle of a size sufficient to deliver an amount of water fog comparable in volume to that of the solid stream.

The master stream is the "big gun" of the fire department. Solid master streams are usually operated at 80 psi (500 kPa); fog master streams at 100 psi (700 kPa). Flows commonly range from 750 to 1,000 gpm (1,650 to 3,750 L/min). They are used when fire streams from handlines would be ineffective. A handline stream may be ineffective because of the magnitude of a fire — when heat is generated faster than it can be absorbed by the stream. Another reason for ineffectiveness may be heat so intense that firefighters

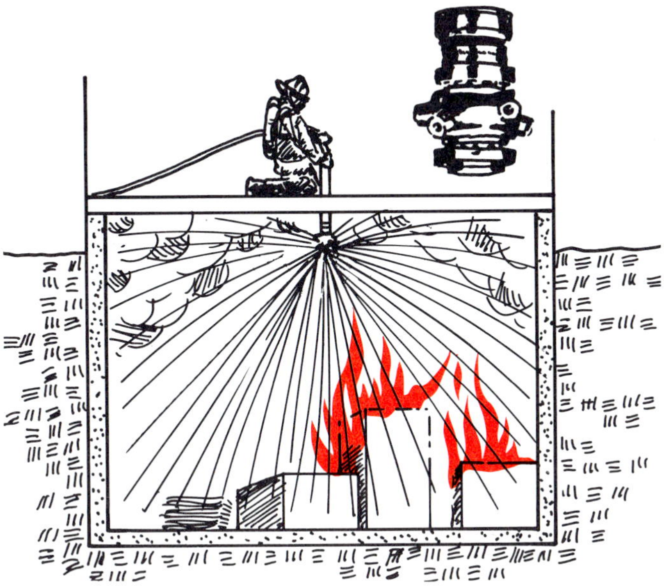

Figure 2.25 — The use of a Bresnan distributor nozzzle for a cellar fire.

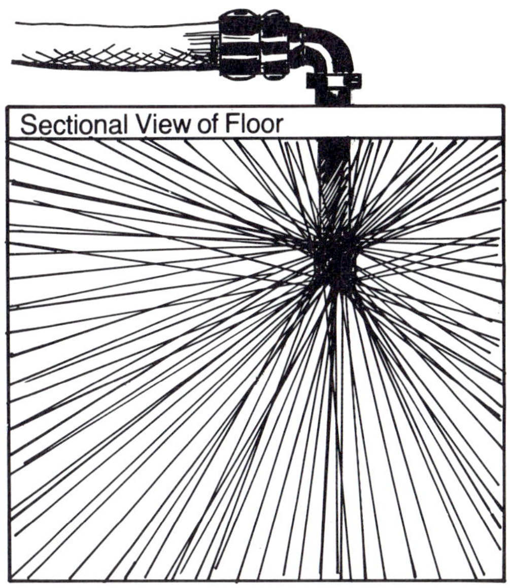

Figure 2.25a — A type of cellar pipe.

Figure 2.25b — A water curtain nozzle in use.

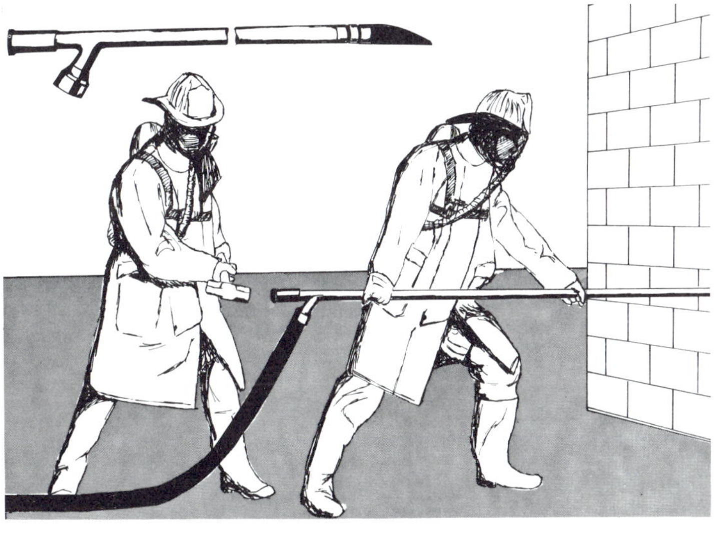

Figure 2.25c — One method of using a piercing nozzle.

cannot approach close enough with handlines. Master streams not only deliver large volumes of water but they have a greater reach than handheld streams. As master stream devices were introduced to the fire service they were identified with a name, a means of recognition, that has earmarked them for their purpose. For many years the term *multiversal* has been applied to several types of master stream devices. This particular term comes from the word *versatile,* which means "capable of turning with ease from one thing to another, changeable, variable and having applications." The combining form *multi* indicates many positions of operation. *This term should not be confused with a particular type of master stream device.*

Figure 2.26 — Use of in-line gauge and pitot gauge to find friction loss of master stream appliances.

Because master stream devices are used from fixed positions most of them have some means for moving the stream in either a vertical or horizontal plane, or both. To permit such adjustments the water must pass through one or more sharp bends near the throat of the nozzle. On some larger master stream devices there are two bends that form a loop in the shape of a ram's horns. This assembly is often referred to as the *ram's horns.* Some other master stream devices have the shape of a lateral, angular *U* (Figure 2.27). The amount of friction loss will vary from device to device (see "Using In-Line Gauges," Chapter 3). Although the friction loss in a master stream device may be assumed to be some fixed figure, such as 10 or 25 psi (70 or 170 kPa), each department must determine the friction loss in the devices it has available. How this may be done with a fair degree of accuracy is shown in Figure 2.26. The difference in the two gauge readings is the friction loss in the device and will vary for each tip size and nozzle pressure.

The effectiveness of any fire stream is tied directly to the mobility of the stream. Much work has been done in recent years to make master streams more mobile. The concept of Large Volume Highly Mobile Mechanical Water Delivery Systems has been introduced. Examples of this concept are the remote-controlled ladder pipes, large nozzles on elevating platforms and the fully controlled nozzle on articulating and telescoping booms.

Turret Pipe

A *turret pipe* is mounted on a deck on fire apparatus and connected directly to the pump (Figure 2.27). The turret pipe is also sometimes called a gun or deck pipe.

Deluge Set

The *deluge set* is composed of a short length of large diameter hose with a large nozzle, supported at the discharge end by a tripod. There is a siamese connection at the supply end (Figure 2.28). The direction and angle of the stream cannot be changed while the deluge set is discharging water.

Many portable monitors are referred to as deluge guns. The difference between these and the deluge set is that the portable monitor uses metal piping instead of hose and the stream direction and angle can be changed while the monitor is discharging water.

Monitor

The *monitor* is a master stream device whose stream direction can be changed while water is being discharged. There are three basic types of monitors: fixed, portable and combination. The fixed monitor is permanently mounted. The portable monitor can be carried to where it is needed. The combination monitor is mounted on the apparatus where it can be used as a turret, but can also be removed and used as a portable monitor wherever needed (Figure 2.31).

The direction and angle of a monitor fire stream can be changed by the use of any of a number of available mechanisms. Depending on the monitor, these mechanisms may incorporate gears, pulleys and chains, electric motors, hydraulics or pneumatics.

Figure 2.27

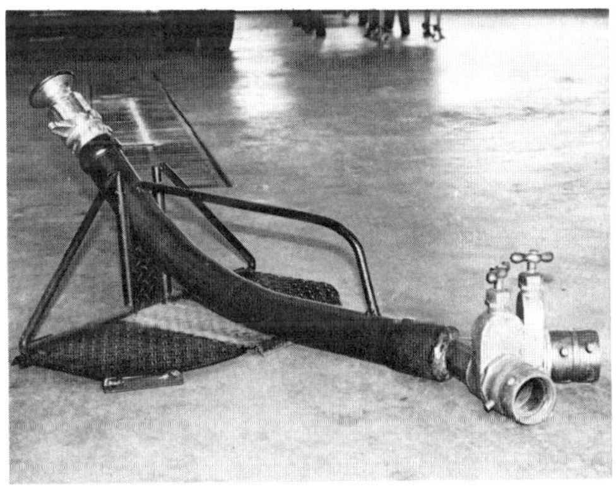

Figure 2.28

Figure 2.27 — Turret pipe

Figure 2.28 — Deluge set

Figure 2.29 — Fixed monitor for industrial system.

Figure 2.30 — Portable monitor.

Figure 2.31 — Combination monitor (mounted on apparatus).

Figure 2.29

Figure 2.30

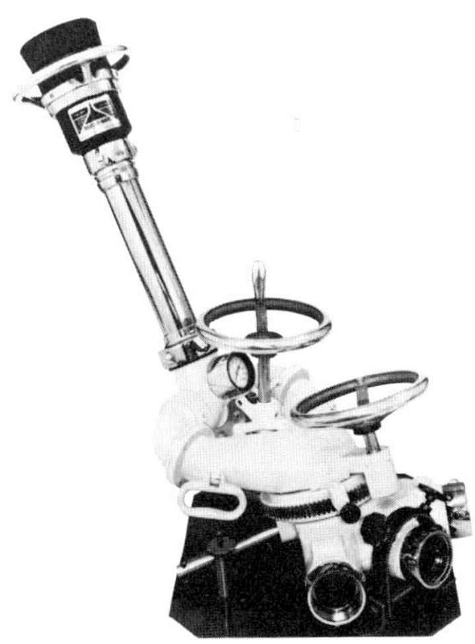

Figure 2.31

Ladder Pipe

A *ladder pipe* is a master stream device attached to the rungs of an aerial ladder (Figure 2.32). The device is supplied by either large diameter hose or metal piping.

Near cousins of the ladder pipe are the streams produced by master stream devices mounted on elevating platforms using articulating or telescoping booms (Figures 2.33 and 2.34). The supply waterways for these apparatus are metal piping that is either articulated or telescoping.

SPECIAL STREAMS

Special stream nozzles are designed for special purposes to meet specific needs. The degree to which such nozzles are needed in the fire service depends upon the particular situation that exists or upon the particular fuel that has set up the situation. The widespread use of hazardous industrial processes, the storage and transportation of flammable liquids, the testing of aviation fuels and engines, and the construction of buildings in general are some of the activities that have increased the demand for special-purpose nozzles for fire fighting. Figures 2.35 - 2.42 show several special-purpose nozzles.

Figure 2.32 — Ladder pipe

Figure 2.33 — Articulated platform.

Figure 2.34 — Telescoping platform.

Figure 2.35 — Cellar pipe and distributor nozzle.

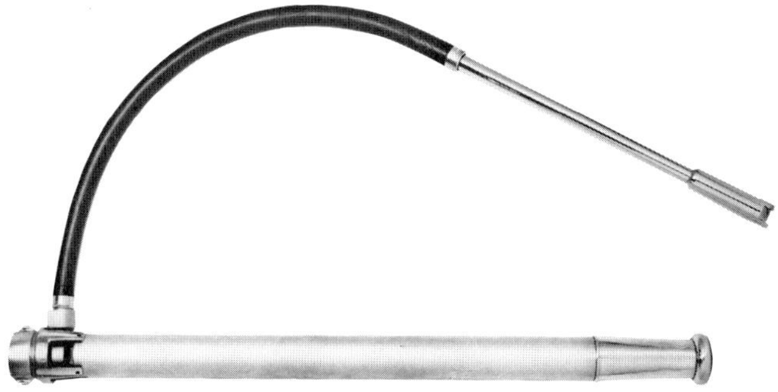

Figure 2.36 — In-line eductor.

Figure 2.37 — Piercing nozzle.

Figure 2.40 Industrial type fog nozzle with rotary shutoff.

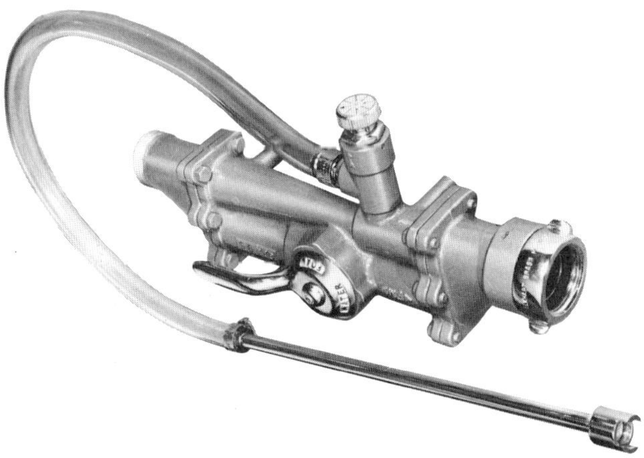

Figure 2.38 — Bypass eductor. Stream can be changed from water to foam or from foam to water without shutting down the line.

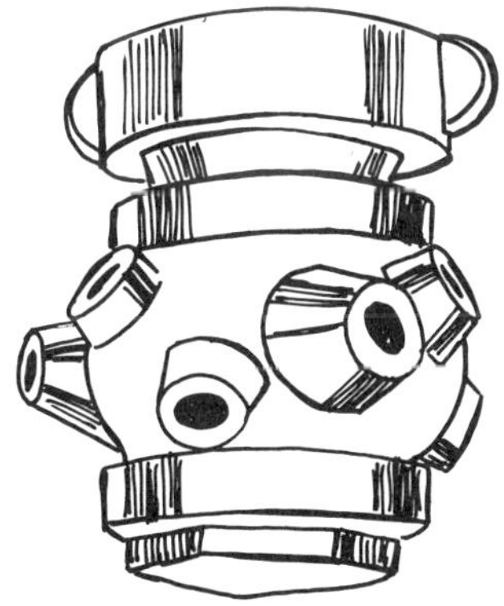

Figure 2.41 — Bresnan Distributor.

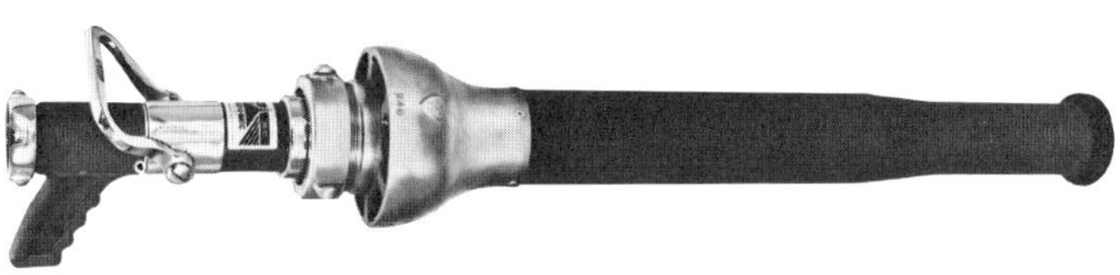

Figure 2.39 — A foam nozzle with shutoff and pistol grip.

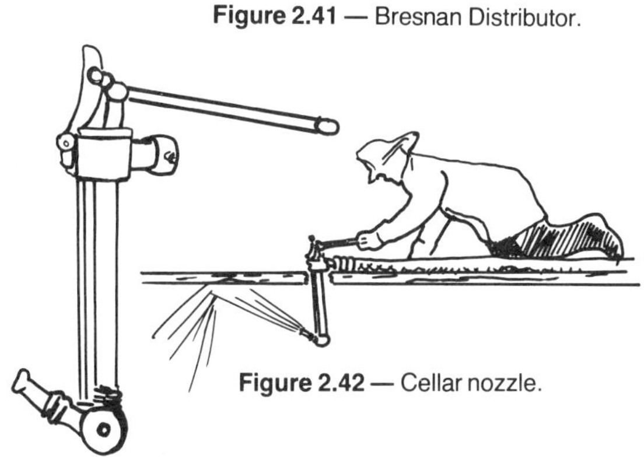

Figure 2.42 — Cellar nozzle.

FOAM

The first foams used in fire fighting were developed in England in the late nineteenth century. These are called *chemical* foams because the bubbles are created by the chemical reactions of mixing together solutions of aluminum sulfate and sodium bicarbonate. The chemical-solution foams were superseded by dry-powder chemical foams, which were an improvement because there were fewer storage and maintenance problems, but there were still drawbacks.

In 1935 the Swedes developed the first liquid foam concentrate for *mechanical* foam (also called airfoam). The bubbles of mechanical foam are created by the mechanical mixing of foam concentrate, water and air, rather than by chemical reaction.

The primary importance of foam is in its ability to extinguish fires for which water alone is inadequate. As is well known, water will sink below the surface of hydrocarbons such as gasoline and oil, and will mix with "polar solvents" such as alcohol and acetone. Foams, however, will float on the surface of flammable liquids. Their extinguishing effect is caused by:

- Smothering: preventing the marriage of air and flammable vapors.
- Suppressing: preventing release of flammable vapors.
- Separating: intervening between the fuel and the fire.
- Cooling: the foam's water content helps to lower the temperature of the fuel and adjacent surfaces.

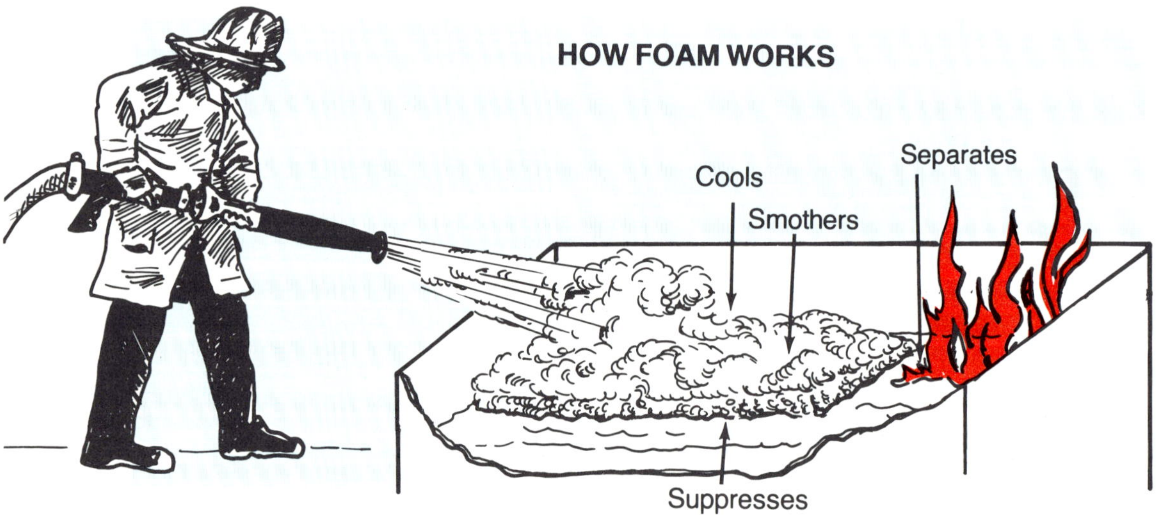

Figure 2.43

There is a variety of foams and foam-producing devices, each of which have their applications, strong points and weaknesses. Although there are "universal" or multi-use foams on the market, there is as yet no one foam that can be used effectively for all types of fires. Each fire department should take inventory of the potential hazards for which foam may be necessary in its area and stock the appropriate types of foam for them.

Proportioning

The proportioner is the device that injects the correct amount of foam concentrate into the water stream to make the foam solution. Correct proportioning is necessary for the production of the optimum quantity and quality of foam. The selection of a proportioner depends on the foam solution flow requirements and, most important, on available water pressure.

The *balanced pressure proportioner* automatically, without manual adjustment, provides the proper concentrate/water solution over a wide range of flows and pressures. These are the most versatile proportioners.

The *pressure proportioner* is another automatic proportioner. It is especially useful when there is low water pressure or there is no electricity for a foam concentrate pump. This type of proportioner and the balanced pressure proportioner are both ideal for foam trucks and fixed systems.

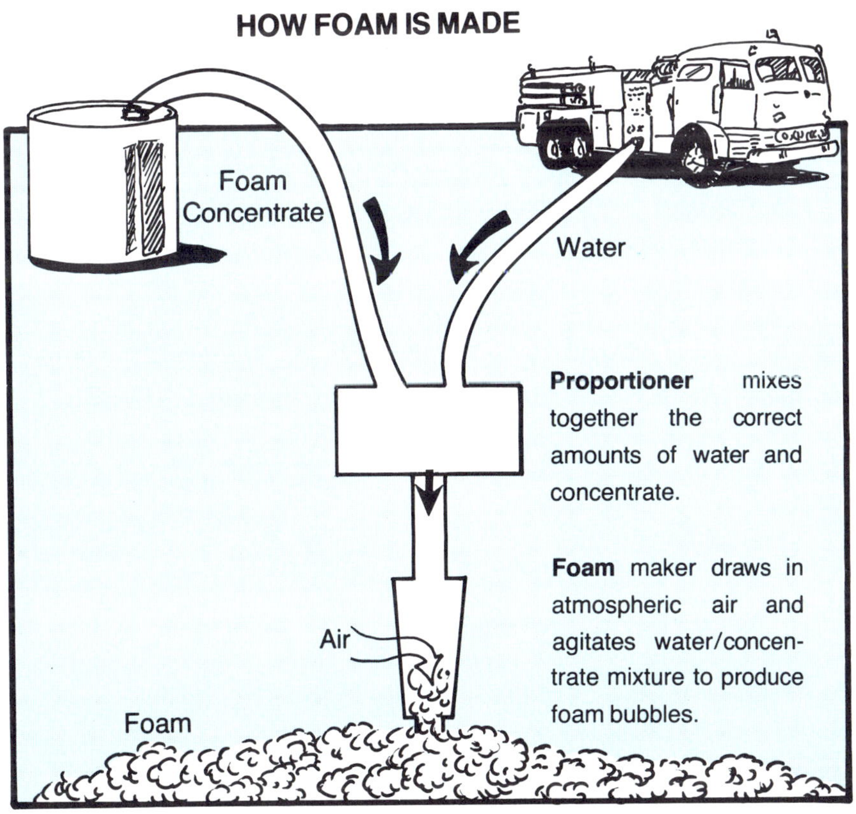

Figure 2.44

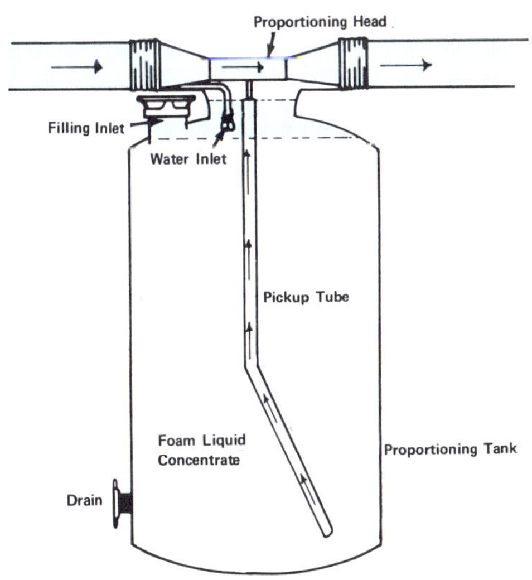

Figure 2.45 — A pressure-proportioning mixing system utilizes the velocity of the water and the density differences of the water and foam concentrate to correctly mix the foam solution.

The *line proportioner* is the simplest and least expensive proportioning device. A line proportioner gives the correct solution ratio only when used with the pressure for which it is designed. A line proportioner works best when the water pressure is reasonably high and it functions properly only with limited back pressure, which puts a limit on the acceptable distance between the eductor and nozzle and on the height to which the nozzle can be elevated.

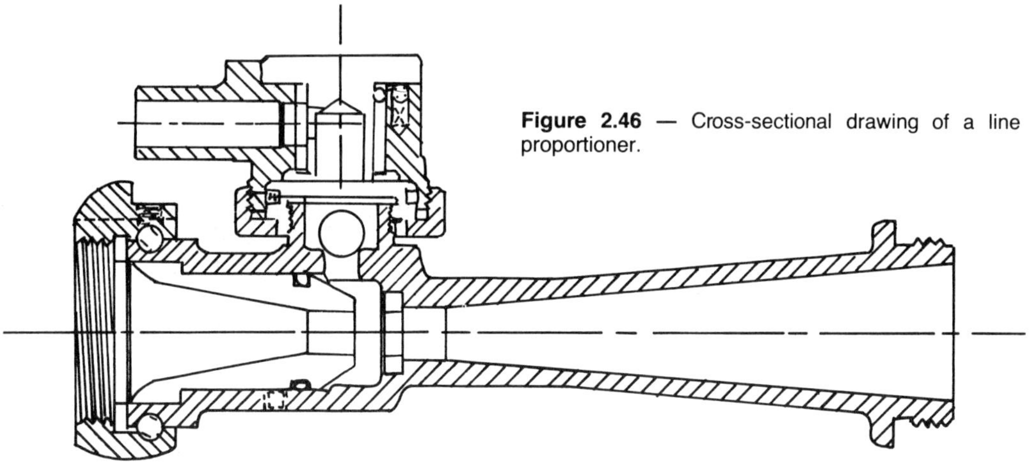

Figure 2.46 — Cross-sectional drawing of a line proportioner.

Foam Makers

After the concentrate and the water have been mixed to make the foam solution, air must be added to make the foam. This is done by the foam maker. Again, there is a variety of foam makers, including foam nozzles, foam chambers, high back pressure foam makers, foam/water sprinklers and special devices. The selection of a foam water depends on the type of foam that is used and on the type of proportioner.

> *Proportioners and foam makers are engineered to work together. Using the wrong proportioner or the wrong foam maker (even if the two are made by the same manufacturer) can result in unsatisfactory foam.*

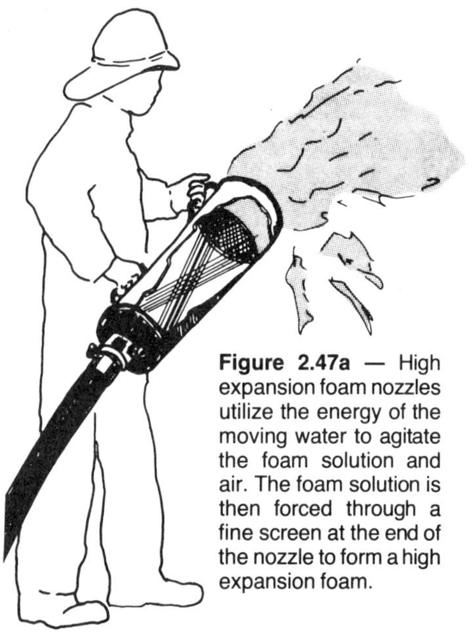

Figure 2.47a — High expansion foam nozzles utilize the energy of the moving water to agitate the foam solution and air. The foam solution is then forced through a fine screen at the end of the nozzle to form a high expansion foam.

Figure 2.47 — Portable high expansion foam generators look similiar to mechanical smoke ejectors. *(Courtesy of Walter Kidde & Co., Inc.).*

Foam makers for low expansion foams use Venturis to draw air into the foam solution. The Venturi sets up sufficient turbulence to create the foam bubbles.

High expansion foam generators produce a high-air-content, stable foam. The air content ranges from 100 parts air to one part foam solution (100×) to 1,000 parts air to one part foam solution (1,000×). There are two basic types of high expansion foam generators: the mechanical blower and the water aspirating. The water aspirating is very similar to other foam-producing nozzles except it is much larger and longer. The back of the nozzle is open to allow airflow. The foam solution is pumped through the nozzle in a fine spray that mixes with air to form a moderate expansion foam. The end of the nozzle has a screen, or series of screens, that breaks the foam up and further mixes it with air. These nozzles typically produce a lower-air-volume foam than do mechanical blower generators.

Mechanical blower generators resemble smoke ejectors in appearance and some can double as smoke ejectors. They operate along the same principle as the water aspirating nozzle except the air is forced through the foam spray instead of being pulled through by the water movement. These devices produce a higher-air-content foam, and are typically associated with total flooding applications. Some manufacturers of smoke ejectors have attachments available that can convert a regular smoke ejector into a high expansion foam generator.

Regular fog nozzles can be used to apply AFFF because the turbulence in the nozzle is sufficient to create AFFF.

Application of Foam

In general, foam should be applied to the fire as gently as possible. This is especially true for alcohol foams. The foam stream should never be applied directly onto the flammable liquid surface unless there is no other approach. The maximum reach of the stream should be used, letting the foam do the work. A foam blanket should be about three inches thick.

Figure 2.48 — Roll foam onto fire from front to prevent spreading the fire.

If the incident is a spill fire on level ground the stream should first be directed to the side, away from the fire, until foam comes from the nozzle. Then the foam stream is directed onto the ground in front of the fire. With a slow sweeping motion the foam blanket is slowly worked onto the fire. The foam stream should never be directed into an existing foam blanket. The force of the stream could blow the existing foam from the surface.

If the fire is in a tank or located so the method above cannot be used, the stream should be directed against the side of the tank or some other object. The tank wall or other object is then a splash board, permitting the foam to slide gently onto and across the liquid surface.

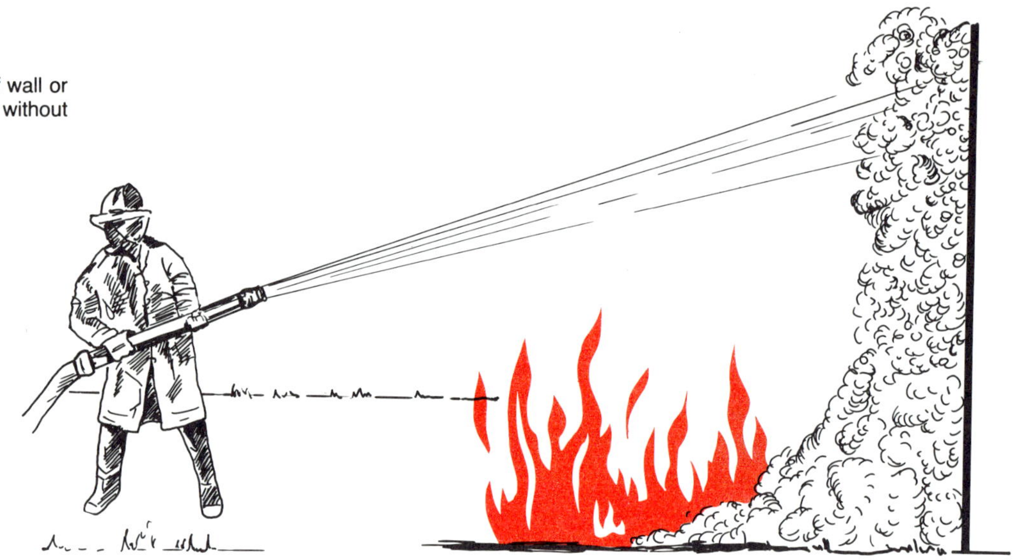

Figure 2.49 — Bank foam off wall or other solid object to cover fire without splattering fuel.

If for any reason foam stops coming from the nozzle the stream should immediately be removed from the foam blanket to prevent washing it away. Operating pressures are very important for the production of quality foam. The acceptable nozzle and proportioner pressures recommended by the foam and device manufacturers should be observed. Overpressures result in a deterioration of quality; underpressures shorten throw and will not produce well expanded foam.

For successful extinguishment the foam must be applied near the recommended rate for the particular foam being used. Lower application rates prolong the time before control and extinguishment, and rates below the critical rate will allow no control of the fire.* The rates of application concern the amount of foam *solution* (not foam) that reaches the fuel surface per minute.

---

*The *minimum recommended rate* of application, generally three times the critical rate, is optimum for practicality of extinguishment and amount of agent required.

TABLE 2.4
Low-Expansion Foam, Minimum Required
(gallons of solution for 10 minutes application)

Nozzle gpm	3%	6%
60	20	40
95	35	60

NOTE: These are minimums for each nozzle. Backup supply should be equal to minimum requirement. With good technique, one-third less AFFF could be used for spill fires.

In general, 5 gallons of 6 percent concentrate will yield 800+ gallons of foam; 5 gallons of 3 percent, 1,600+ gallons of foam.

If the foam nozzle being used has a spray stream attachment the attachment should be used because it permits gentler application, less mixing of fuel and foam, and (especially for spill fires) can provide some radiant-heat protection for firefighters and exposures.

There are several cautions that must be remembered when foam is being used. Regular water streams can be used to cool adjacent materials or to protect exposures, but should never be used in such a way as to disturb the foam blanket. Foam streams are more electrically conductive than is regular water — even as a fog stream there is great potential danger, so foam should never be used on energized electrical equipment. Entraining contaminated air (dust, products of combustion) should be avoided — it has negligible effect on low expansion foam, but can reduce the quality of high expansion foam. Foam should not be used on materials that are stored as liquids but that are gaseous at ambient temperatures and pressure (propane, butadiene, vinylchloride, for example). Remember that foam really is just bubbly water — don't use it on materials that react with water.

Foam conducts electricity

All of the above procedures and cautions are necessarily general, applicable to most foams. For more complete information consult NFPA Standard Nos. 11, 11A and 11B. Always follow manufacturers' recommendations.

Low Expansion Foams: Protein

Low expansion foams are so named to differentiate them from high expansion foams. The basic differences are that low expansion foams have a smaller air/solution ratio and their primary use is for flammable liquid fires. Low expansion foams can be used to good effect when the temperature of the bulk of a contained liquid does not exceed 250°F. The foams can be used on spill fires (depth less than one inch) with higher product temperatures; there is no danger of boilover.

REGULAR PROTEIN FOAMS

Regular protein foams are chemically broken-down natural protein solids. The end product of this chemical digestion is protein polymers, long strands, that have excellent elasticity, water retention capabilities and high strength. These characteristics make protein foam excellent for use as a fire fighting agent. Basic protein foams are nontoxic and work within a 100°F temperature range from 20° to 120°F, but, as with all foams, low water temperatures help to produce a more stable foam. The preferred water temperature range for low expansion foams such as this is 35° to 85°F. Either fresh or sea water can be used. These foams are not well-suited for use on polar solvents, for extremely cold temperatures, for subsurface injection, or for use with dry-chemical powders. However, certain organic solvents can be added to the foam to improve the suitability for use in these situations.

Low water temperatures give more stable foam

Regular protein foams, as well as most other foams, are marketed to be mixed into 3 percent or 6 percent foam solutions. This means that in a 3 percent foam solution for every one hundred parts of foam solution (foam concentrate and water mixed) there are 3 parts of concentrate to 97 parts water. For 6 percent solution there are 6 parts concentrate and 94 parts water. The main reason for this difference in concentration is because of technological advancements in foam concentrate production. Initially, the best concentrate available was 6 percent. After a time, procedures were improved to allow a suitable 3 percent foam concentrate to be marketed. The 3 percent gave the advantage of costing less to ship per area of coverage, and requiring only one-half the storage space of 6 percent foam concentrates. Six percent foam concentrates are still marketed to meet the needs of departments that cannot afford to convert their systems to a 3 percent foam concentrate.

Low-temperature foams are protein-based foams with nonflammable antifreeze solutions added to them. These foams can be used at air temperatures down to −20°F. As with normal protein foam, they are marketed in 3 percent and 6 percent solutions.

Regular protein foams have many positive qualities. Their adhesiveness gives them the ability to cover vertical surfaces, which helps to cool tank walls and automobile and aircraft bodies, and their relative stiffness gives greater range from portable devices. They have high burnback resistance so the foam will adhere even to hot metal, forming a seal. With these qualities, low expansion foams are ideally suited for application through portable devices, for gasoline and oil fires causing large amounts of hot metal, and for vehicular fires where the insulation of exposures is important.

FLUOROPROTEIN FOAMS

Fluoroprotein foams are basically protein foams fortified with fluorinated solvents. The addition of these solvents give fluoroprotein foams several advantages over regular protein foams. One of the primary advantages is that these foams tend to separate from the flammable liquids into which they are mixed. This quality makes fluoroprotein foams excellent for subsurface injection, or surface applications in which the foam becomes agitated with the flammable liquid. This quality also enhances the foam's ability to suppress vapor release even when the foam is saturated with the fuel. Other advantages are that fluoroprotein foams tend to be more stable than regular protein foams and more compatible for use with dry-chemical agents.

Fluoroprotein excellent for subsurface injection

Fluoroproteins have the same temperature characteristics as regular protein foams. It is also marketed in 3 percent and 6 percent concentrates.

ALCOHOL FOAMS

Alcohol foams have been developed for use on polar solvents such as alcohol, lacquer thinner, acetone and ketones. Polar solvents are miscible in water. This characteristic causes regular foams to break down rapidly. The alcohol foams were developed because regular foams are very miscible in polar solvents, and tended to melt into the burning liquid without extinguishing the fire. Additionally, regular hydrocarbon liquids mixed with even small amounts of polar solvents tended to destroy the effectiveness of regular foam products.

There is one type of protein-based alcohol foam. This foam is derived from regular protein foam mixed with heavy-metal salts suspended in organic solvents. Protein-based alcohol foams must, to be effective, be applied gently to the burning surface — the least submergence of the foam will destroy it — and must be applied immediately after eduction into the water. Protein-based alcohol foams when pumped through hose or piping lose effectiveness. This loss is in proportion to the distance it is pumped. For this reason, protein-based alcohol foams are usually mixed into the water stream at or near the application nozzle. Alcohol foams can be used on hydrocarbon fires, but with less effectiveness than regular protein or fluoroprotein foam. Application temperature is from 35° to 120°F.

Apply protein-based alcohol foams gently

Low Expansion Foams: Synthetic

ALCOHOL FOAMS

There are two types of synthetic alcohol foams. Both of these foams have the advantage over protein-based alcohol foams in that they do not have to be applied gently to the surface, and can be pumped long distances without loss of effectiveness. Both are marketed in 3 percent and 6 percent solutions and have the same temperature range as protein alcohol foam.

The first type of synthetic alcohol foam is catalytic alcohol foam. It is the result of the mixing of a two-part solution, a polymer and catalyst. The resulting foam is a very stable alcohol-resistant foam. This foam can be pumped long distances without loss of effectiveness.

The second synthetic alcohol foam is sometimes referred to as multi-use foam. This is because it can be used effectively on either hydrocarbon or polar liquids and work effectively when applied through most foam-mixing devices. These foams are synthetically produced concentrates in a single-component foam. The proportioning is very similar to that of regular protein foams, and they may be pumped long distances without loss of effectiveness.

SYNTHETIC DETERGENT FOAMS

Several synthetically produced detergent foams are available. These foams have a high foam yield, but are generally less stable than other foams. They have a low resistance to heat or physical destruction, and require high application rates to be effective.

Detergent foams cause a very low surface tension of water, and may be thought of as wetting agents. This makes their use on Class A combustibles effective and will sometimes form a frothy emulsion on top of flammable liquids. However, this characteristic causes detergent foams to break down other foams.

AQUEOUS FILM FORMING FOAMS

Aqueous film forming foams (AFFF) are dual-action synthetic foams. The first action of AFFF is similar to the detergent foam's air-entrapping action in forming a blanket of strong foam that spreads over the burning liquid surface. This blanket smothers

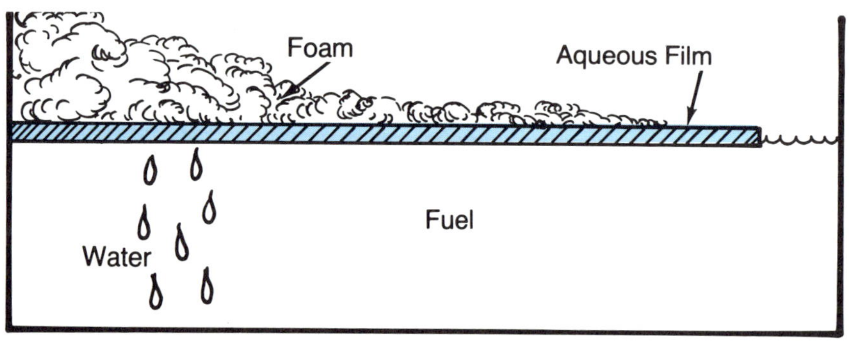

Figure 2.50

the fire and retards vaporization of the flammable liquid to below flammable limits. The second action is controlled solution drainage from the foam blanket, which forms an aqueous solution film without bubbles between the fuel surface and the blanket. This aqueous film smothers the fire and prevents vaporization just as the foam film does. The aqueous film is self-healing, and will re-cover open areas caused by agitation of the flammable liquid surface. The result of the foam action and the film forming action make AFFF one of the most dependable and versatile foams available. Use of AFFF has the added advantages of not requiring special application devices being compatible with dry chemical, and gives comparatively quick knockdown.

AFFF is self-healing

AFFFs are generally strengthened by the addition of fluorinated solvents like those used to stabilize protein foams. These foams can be used on flammable liquids, and under some conditions on polar solvents. Often AFFF films are used on flammable liquid spills to prevent ignition of the flammable liquid. AFFFs are marketed in 3 percent and 6 percent solutions, and have temperature characteristics like alcohol foams.

HIGH EXPANSION FOAMS

High expansion foams are foam concentrates mixed into a 2 percent solution and then mixed with air to form high-air-content, good-quality foams. They can be used for Class A fires, especially indoors or where wind will not disrupt the blanket. They can also be used for Class B fires (see NFPA Standard No. 11A, section 1-4e). They are ineffective on polar solvent fires.

High expansion foams can be thought of as three-dimensional water. Their ability to totally fill large areas has made them quite suitable for fixed installations in such places as aircraft hangars, but they also have many applications through portable devices.

High expansion foams can be very effective for fires in confined areas such as basements, mine shafts, sewers and concealed spaces. They can also be used to prevent fire spread by blocking concealed spaces, ducts and areas of vertical spread such as stairwells and elevator shafts, and by insulating exposures.

Ventilation is very important when high expansion foams are used. There must be some outlet for the air displaced by the foam in a confined space or the foam will not fill the area. The lightness of the foam gives it little range, so firefighters must be rather close to the fire and attack from the windward side.

Ventilation important with high expansion foam

3

HYDRAULICS CALCULATIONS

NFPA STANDARD 1002

3–4 Hydraulic Calculations.

3–4.1 The fire apparatus driver/operator shall demonstrate the principles of friction loss as they apply to: . . .

(f) Use of appliances

(g) Use of multiple hose lines

3–4.3 The fire apparatus driver/operator shall identify the following terms that relate to the basic principles of fire service hydraulics . . .

(f) Head pressure (back pressure) . . .

(h) Net engine pressure

3–4.5 The fire apparatus driver/operator shall identify and demonstrate the determination and use of square roots as required to solve fire department pumper hydraulic problems.

3–4.6 The fire apparatus driver/operator shall identify and demonstrate the use of fractions, percentages, and decimal fractions in the mathematical calculations as required to solve fire department pumper hydraulic problems.

3–4.7 The fire apparatus driver/operator shall demonstrate the use of simple algebraic formulas required to solve fire department pumper hydraulic problems.

3–4.8 The fire apparatus driver/operator, given a series of fire ground situations and using the written formulas specified by the authority having jurisdiction, shall determine:

(a) Nozzle or pump discharge pressure when the length and size of hose, and size of nozzle are given.

(b) Water flow in gallons per minute when the diameter of the orifice and pressure at the orifice are given.

(c) The friction loss in the supply and attack lines, used by the authority having jurisdiction, when the GPM flow is given.

(d) Friction loss in siamesed lines when size of hose and GPM flow are given.

(e) Friction loss in wyed lines when size of hose and GPM flow are given.

(f) Friction loss in multiple lines when the size of hose and GPM flow are given.

(g) An estimated remaining available volume from a hydrant while pumping a given volume.

3–4.9 The fire apparatus driver/operator, given a series of fire ground situations, shall calculate correct pump discharge pressure, GPM, friction loss, and nozzle pressure, using mental formulas specified by the authority having jurisdiction.

3–4.10 The fire apparatus driver/operator, given a series of fire ground situations involving various operating pressures, shall demonstrate the formula for calculation of nozzle reaction of hand and master streams used by the authority having jurisdiction.

3–4.11 The fire apparatus driver/operator, given the necessary information, shall compute the maximum lift of a fire department pumper.

FORMULA SYMBOLS

Equations such as the algebraic expressions of laws and rules are called *formulas.* They are more easily remembered and applied than worded rules; consequently, they are commonly used by the technician and engineer. Custom has developed the use of certain letters to represent mathematical terms. Some of these, however, are not standard. Following are the common letters and abbreviations used as symbols that are used in this manual.

EL	elevation loss	FL	friction loss
r	radius	V	volume or velocity
A	area	P	flow pressure
d	diameter	gpm	gallons per minute
π	a constant (3.1416)	psia	pounds per square inch absolute
$\sqrt{\ }$	square root	psig	pounds per square inch gauge
g	gravity (acceleration)	s	distance
h	height	t	time
L	length	PDP	pump discharge pressure
AL	appliance loss	NP	nozzle pressure

SQUARE ROOT

The square root of a number is a number that when multiplied by itself equals the original number. For example, the square root of 64 is 8, because 8(8) = 64. The square root of 81 is 9, because 9(9)= 81. To avoid writing out the words *square root in value,* the sign $\sqrt{\ }$ is used. This means that the square root of the number under the sign is to be taken.

SQUARING A NUMBER

The opposite of square root is squaring a number. This means a certain number is multiplied by itself (for example, 8(8) = 64). To avoid writing 8(8) the number 2 is placed above and to the right of the eight as a superscript to indicate that 8 is multiplied by itself (for example, 8^2).

Frequently, when computing friction losses and fire flow a firefighter must square a number. Many of these numbers end in 5 or in a fraction whose decimal equivalent ends in 5, such as the following:

½	¾	1¼	1½	1¾	2½	4½
0.5	0.75	1.25	1.5	1.75	2.5	4.5

A rapid method that may be used to square a number ending in 5 is as follows: First square the digit 5 and write down the resulting 25. Then multiply the other number by the number that is one larger than itself and write the product to the left of the 25.

TABLE 3.1

SQUARE ROOTS OF NUMBERS 1-250									
n	$\sqrt{n}$	n	$\sqrt{n}$	n	$\sqrt{n}$	n	$\sqrt{n}$	n	$\sqrt{n}$
1	1.	51	7.1414	101	10.0499	151	12.2882	201	14.1774
2	1.414	52	7.2111	102	10.0995	152	12.3288	202	14.2127
3	1.732	53	7.2801	103	10.1489	153	12.3693	203	14.2478
4	2.000	54	7.3485	104	10.1980	154	12.4097	204	14.2829
5	2.236	55	7.4163	105	10.2470	155	12.4499	205	14.3178
6	2.449	56	7.4833	106	10.2956	156	12.4900	206	14.3527
7	2.646	57	7.5498	107	10.3441	157	12.5300	207	14.3875
8	2.828	58	7.6158	108	10.3923	158	12.5698	208	14.4222
9	3.000	59	7.6811	109	10.4403	159	12.6095	209	14.4568
10	3.162	60	7.7460	110	10.4881	160	12.6491	210	14.4914
11	3.3166	61	7.8102	111	10.5357	161	12.6886	211	14.5258
12	3.4641	62	7.8740	112	10.5830	162	12.7279	212	14.5602
13	3.6056	63	7.9373	113	10.6301	163	12.7671	213	14.5945
14	3.7417	64	8.0000	114	10.6771	164	12.8062	214	14.6287
15	3.8730	65	8.0623	115	10.7238	165	12.8452	215	14.6629
16	4.0000	66	8.1240	116	10.7703	166	12.8841	216	14.6969
17	4.1231	67	8.1854	117	10.8167	167	12.9228	217	14.7309
18	4.2426	68	8.2462	118	10.8628	168	12.9615	218	14.7648
19	4.3589	69	8.3066	119	10.9087	169	13.0000	219	14.7986
20	4.4721	70	8.3666	120	10.9545	170	13.0384	220	14.8324
21	4.5826	71	8.4261	121	11.0000	171	13.0767	221	14.8661
22	4.6904	72	8.4853	122	11.0454	172	13.1149	222	14.8997
23	4.7958	73	8.5440	123	11.0905	173	13.1529	223	14.9332
24	4.8990	74	8.6023	124	11.1355	174	13.1909	224	14.9666
25	5.0000	75	8.6603	125	11.1803	175	13.2288	225	15.0000
26	5.0990	76	8.7178	126	11.2250	176	13.2665	226	15.0333
27	5.1962	77	8.7750	127	11.2694	177	13.3041	227	15.0665
28	5.2915	78	8.8318	128	11.3137	178	13.3417	228	15.0997
29	5.3852	79	8.8882	129	11.3578	179	13.3791	229	15.1327
30	5.4772	80	8.9443	130	11.4018	180	13.4164	230	15.1658
31	5.5678	81	9.0000	131	11.4455	181	13.4536	231	15.1987
32	5.6569	82	9.0554	132	11.4891	182	13.4907	232	15.2315
33	5.7446	83	9.1104	133	11.5326	183	13.5277	233	15.2643
34	5.8310	84	9.1652	134	11.5758	184	13.5647	234	15.2971
35	5.9161	85	9.2195	135	11.6190	185	13.6015	235	15.3297
36	6.0000	86	9.2736	136	11.6619	186	13.6382	236	15.3623
37	6.0828	87	9.3274	137	11.7047	187	13.6748	237	15.3948
38	6.1644	88	9.3808	138	11.7473	188	13.7113	238	15.4272
39	6.2450	89	9.4340	139	11.7898	189	13.7477	239	15.4596
40	6.3246	90	9.4868	140	11.8322	190	13.7840	240	15.4919
41	6.4031	91	9.5394	141	11.8743	191	13.8203	241	15.5242
42	6.4807	92	9.5917	142	11.9164	192	13.8564	242	15.5563
43	6.5574	93	9.6437	143	11.9583	193	13.8924	243	15.5885
44	6.6332	94	9.6954	144	12.0000	194	13.9284	244	15.6205
45	6.7082	95	9.7468	145	12.0416	195	13.9642	245	15.6525
46	6.7823	96	9.7980	146	12.0830	196	14.0000	246	15.6844
47	6.8557	97	9.8489	147	12.1244	197	14.0357	247	15.7162
48	6.9282	98	9.8995	148	12.1655	198	14.0712	248	15.7480
49	7.0000	99	9.9499	149	12.2066	199	14.1067	249	15.7797
50	7.0711	100	10.0000	150	12.2474	200	14.1421	250	15.8114

Example 1

To square the number 25:

1. Square the digit 5. $5^2 = 25$. Write down 25.
2. Multiply the 2 by the next number larger. $2(3) = 6$.
3. Write down 6 to the left of the 25, and the result (the square of 25) is 625.

Example 2

If the number to be squared is 2½, first turn the ½ into its decimal equivalent, 0.5. After that the process is the same except that the decimal point is placed two places to the left — 6.25.

Example 3

To square 1.25 the process is the same:

1. Square the digit 5. $5^2 = 25$. Write down 25.
2. Multiply 12 by the next number larger. $12(13) = 156$.
3. Write down 156 to the left of the 25. Move the decimal points four places to the left. The square of 1.25 is 1.5625.

Space provided for calculations and notes on pages 196 and 197.

Exercise 1

What is the square of 3½? (Answers to all Chapter 3 exercises are at the end of the chapter.)

Exercise 2

What is the square of 7.25?

Exercise 3

What is the square of 55?

USING CONSTANTS IN FORMULAS

In solving formulas there generally appear numbers that are definite and always the same. Such numbers are known as *constants*.

One example of a constant factor is *pi*, the symbol for which is π, in the measurement of circles. It always has the same value — 3.1416 — so it is a constant. The formula for finding the area of a circle, for example, is $A = \pi r^2$.

Example 4

Find the area of a nozzle whose diameter is 2½ inches. (To keep from having to halve 2.5 to find the radius, the formula below may be used.)

$$A = \frac{\pi d^2}{4}$$

$A = \pi r^2$	$A = \frac{\pi d^2}{4}$
$A = 3.1416(1.25^2)$	$A = \frac{3.1416(2.5^2)}{4}$
$A = 3.1416(1.56)$	$A = \frac{3.1416(6.25)}{4}$
$A = 4.9$	$A = \frac{19.635}{4} = 4.9$

Exercise 4

What is the area of a nozzle whose diameter is 1½ inches?

Exercise 5

What is the area of the surface of the water in a storage tank with a diameter of 30 feet?

CAPACITY OF A CYLINDRICAL TANK

A short-cut formula that may be used in the field for calculating the approximate capacity in gallons of any large cylindrical tank is the diameter squared times the constant 6 times the length of the tank:

$$gallons = d^2 6L$$

Example 5

To find the approximate capacity of a cylindrical tank 10 feet in diameter and 15 feet long:

$$gallons = 10^2(6)(15)$$

$$gallons = 100(6)(15)$$

$$gallons = 600(15)$$

$$gallons = 9{,}000$$

This figure is sufficiently accurate for field estimates. The use of this short-cut formula will permit calculations to be made when they otherwise might be neglected.

A longer, but accurate, method is to first find the number of cubic feet in the tank, using the formula:

$$ft^3 = \pi r^2 L$$

Example 6

$$ft^3 = 3.1416(5^2)(15)$$
$$ft^3 = 3.1416(25)(15)$$
$$ft^3 = 78.54(15)$$
$$ft^3 = 1{,}178$$

There are 7.48 U.S. gallons in a cubic foot, so the tank has a capacity of 7.48(1.178) = 8,811 gallons.

The same formula can be used for SI units*

$$m^3 = \pi r^2 L$$
$$m^3 = 3.1416(1.524^2)(4.572)$$
$$m^3 = 3.1416(2.323)(4.572)$$
$$m^3 = 33.46$$

1 m^3 = 1 kl, so the capacity of the tank is 33.46 kl, or 33,460 liters (which, converted, is 8,840 gallons).

---

*Not all formulas that work for English units will work for SI units. For example, the short-cut formula will not work for SI, as is shown below:

$$\text{liters} = d^2 6L$$
$$\text{liters} = 3.04^2 6(4.572)$$
$$\text{liters} = 9.24(6)(4.572)$$
$$\text{liters} = 55.44(4.572)$$
$$\text{liters} = 253.5$$
$$253.5 \text{ liters} = 66.97 \text{ gallons}$$

This, of course, does not mean that the short-cut method is worthless; only that a direct "plugging in" of comparable SI units will not work for formulas such as this that are derived specifically for English units of measurement.

Find the answers to Exercises 6 and 7 by using both the short-cut and the exact formulas.

Exercise 6

What is the capacity of a cylindrical tank 30 feet long and 10 feet in diameter?

Exercise 7

What is the capacity of a cylindrical tank 10 feet long and 3 feet in diameter?

COMPUTING WATERFLOW IN GALLONS PER MINUTE

The easiest way to determine how much water is flowing from the orifice of a nozzle is to refer to such prepared tables as Table A.4. These tables have been computed by using a formula for gallons per minute flow when the flow pressure is known. The formula may be stated "Gallons per minute flow is equal to 29.7 times the diameter of the orifice squared times the square root of the flow pressure." The formula is written

$$gpm = 29.7d^2\sqrt{P}$$

where

d = orifice diameter (inches)

P = psi (Pitot tube and gauge reading)

29.7 = a constant (30 may be used for rough estimates and field work)

Example 7

If one assumes that a 1-inch nozzle tip is used and the flow pressure is 50 psi, the formula would read

$$gpm = 29.7(1^2)(7.0711)$$

$$gpm = 29.7(7.0711)$$

$$gpm = 210$$

Exercise 8

A master stream device with a 2-inch tip is operating at 80 psi. What is the gallons per minute flow?

Exercise 9

What would be the gallons per minute flow from a $1\frac{1}{8}$-inch tip at 60 psi?

FRICTION LOSS

To produce effective fire streams the amount of friction loss in the fire hose between the pumper and the nozzle must be known. Friction loss is caused by a number of factors, such as hose and coupling condition, kinks and bends, as discussed earlier in this manual. The primary determinant, however, is the volume of water flowing per minute, taking into account the length and diameter of the hoseline. Since the amount of hose that may be used between the pumper and the nozzle will not always be the same, it is most convenient to consider friction loss in terms of pounds per square inch (psi) per 100 feet of hose (or kiloPascals [kPa] per 30 m of hose).

Manufacturers are constructing fire hose of materials other than cotton and rubber. These materials not only provide a smoother surface on the inner tube lining but also make possible larger inside diameters without increasing weight or bulk. The friction loss of such hose varies and friction-loss formulas, tables or rule-of-thumb methods of determining pump and nozzle pressures will give approximate results. (The formula $2Q^2$ is sometimes used, which may yield somewhat more correct answers for 2½-inch hose friction loss.) The friction-loss formulas and tables given in this manual were developed and are useful for cotton-jacketed, rubber-lined hose. Synthetic hose can have different friction losses because of less or greater expansion, smoother lining, etc., and should be tested to determine the actual friction loss.

Although friction loss can be *measured* with considerable accuracy, the results may not be consistent with fire stream calculations. There is only one sure way to know the friction losses of a layout. Hose should be "laid out" and accurate results of friction losses recorded when various volumes are flowing. This enables firefighters to determine friction losses in a particular type of hose, compute pump discharge pressures and establish their own departmental policy for nozzle and pump discharge pressures at fires.

Some research has been done in the reduction of friction loss through the use of an additive, a linear polymer that forms a straight-line chain of water molecules and reduces turbulence and frictional drag. With the idea of flowing larger volumes of water through currently used hose diameters, or being able to reduce hose diameters to carry the same amount of water as larger hoses, some fire departments have been experimenting with the agent and various means of injecting it into the waterflow. Some tests using the additive have shown an increase of about 40 percent more water with the same friction loss or a reduction of 40 percent in friction loss at the same flow. The additive also increases the reach of a solid stream and appears to make fog streams denser.

2½-Inch Hose

Instead of using *V* for volume, the letter *Q* is commonly used to signify the waterflow. Quantity *(Q)* is stated in *the number of hundreds* of gallons per minute *(gpm)* flow. The quantity in hundreds is used in a formula to compute the friction loss in cotton-jacketed, rubber-lined 2½-inch hose. The following formula is used to find *Q*:

$$Q = \frac{gpm}{100}$$

After *Q* has been determined the following formula can be used to calculate friction loss in 100 feet of 2½-inch hose when more than 100 gpm is flowing.

$$FL = 2Q^2 + Q$$

where

FL = friction loss

Q = hundreds of gpm

Example 8

If 200 gpm is flowing from a nozzle the friction loss for 100 feet is calculated as follows:

$$Q = \frac{gpm}{100}$$

$$Q = \frac{gpm}{200}$$

$$Q = 2$$

$$FL = 2Q^2 + Q$$

$$FL = 2(2^2) + 2$$

$$FL = 2(4) + 2$$

$$FL = 8 + 2$$

FL = 10 psi per 100 feet of 2½-inch hose

Exercise 10

If 250 gpm is flowing, what will be the friction loss per 100 feet of 2½-inch hose?

Exercise 11

Find the friction loss per 100 feet of 2½-inch hose when there are 400 gpm flowing.

Exercise 12

What is the total friction loss in 500 feet of 2½-inch hose when there are 300 gpm flowing?

Exercise 13

Find the total friction loss in 400 feet of 2½-inch hose equipped with a 1-inch tip flowing at 65 psi.

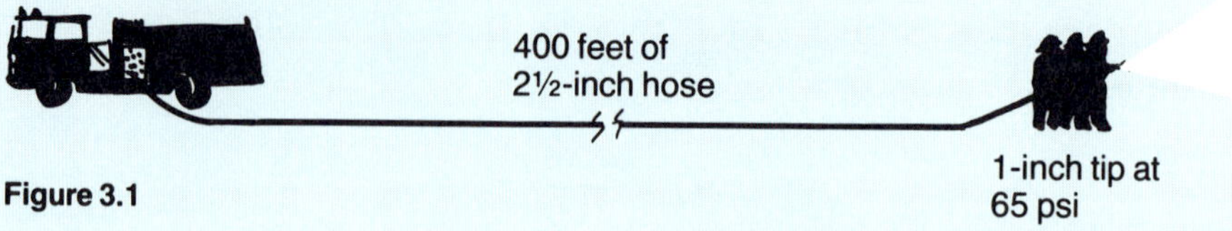

Figure 3.1

Flows of 100 GPM or Less

When flows from a nozzle or nozzles are near to or less than 100 gpm, the following formula should be used for 2½-inch hose.

$$FL = 2Q^2 + \tfrac{1}{2}Q$$

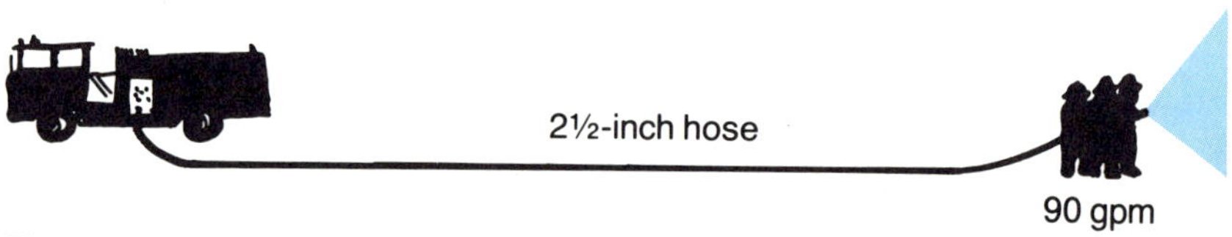

Figure 3.2

Example 9

The friction loss per 100 feet of 2½-inch hose with a flow of 90 gpm from the nozzle is calculated as follows:

$$Q = \frac{gpm}{100}$$

$$Q = \frac{90}{100}$$

$$Q = 0.9$$

$$FL = 2Q^2 + \tfrac{1}{2}Q$$

$$FL = 2(0.9^2) + 0.5(0.9)$$

$$FL = 2(0.81) + (0.45)$$

$$FL = 1.62 + 0.45$$

$FL = 2.07$ psi per 100 feet of 2½-inch hose

Other than 2½-Inch Hose

Friction loss for hose larger or smaller in diameter than 2½ inches can be determined by first finding what the friction loss would be in 2½-inch hose with the same gpm flowing (that is, to find the friction loss for 3-inch hose flowing 200 gpm, for example, first find the friction loss for 2½-inch hose flowing 200 gpm). The 2½-inch hose friction loss is then either multiplied or divided by the appropriate conversion factor (Table 3.2). Whether multiplication or division is used is up to personal preference.

TABLE 3.2
Conversion Factors for Finding Friction Loss

Diameter (in inches) and Type (single lines)	Divide by	Multiply by
¾ booster	0.0029	344.00
1 booster	0.011	91.00
1¼ rubber lined	0.025	40.00
1½ rubber lined	0.074	13.50
1¾ rubber lined	0.16	5.95
2 rubber lined	0.34	2.94
3 rubber lined	2.60	0.385
3 rubber lined with 2½-inch couplings	2.50	0.40
3½ rubber lined	5.80	0.172
4 rubber lined	11.00	0.09
4½ rubber lined	19.50	0.051
5 rubber lined	32.00	0.031
6 rubber lined	83.33	0.012
1¼ unlined linen	0.0157	63.60
1½ unlined linen	0.039	25.60
2 unlined linen	0.16	6.25
2½ unlined linen	0.47	2.13

Siamesed Lines of Equal Length	Divide by	Multiply by
Two 2½-inch	3.60	0.28
Three 2½-inch	7.75	0.129
Two 3-inch	9.35	0.107
One 3-inch, One 2½-inch	6.10	0.164
Two 2½-inch, One 3-inch	11.50	0.087
Two 3-inch, One 2½-inch	15.00	0.067

Standpipes (diameter)	Divide by	Multiply by
4 inches	7.50	0.133
5 inches	22.00	0.045
6 inches	52.00	0.019

Example 10

What will be the friction loss per 100 feet of 1½-inch hose where there are 60 gpm flowing?

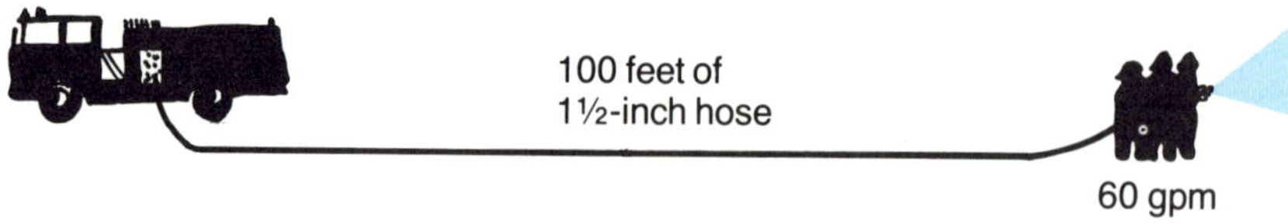

Figure 3.3

$$Q = \frac{gpm}{100}$$

$$Q = \frac{60}{100}$$

$$Q = 0.60$$

$$FL = 2Q^2 + \tfrac{1}{2}Q$$
$$FL = 2(0.36) + 0.30$$
$$FL = 0.72 + 0.30$$
$$FL = 1.02 \text{ psi per 100 feet of 2½-inch hose}$$

Conversion factors:

Multiply by 13.5
or
Divide by 0.074

$$1.02(13.5) = 13.77 \text{ psi}$$
or
$$\frac{1.02}{0.074} = 13.7 \text{ psi}$$

Example 11

What will be the friction loss per 100 feet of 3-inch hose with 2½-inch couplings when there are 500 gpm flowing?

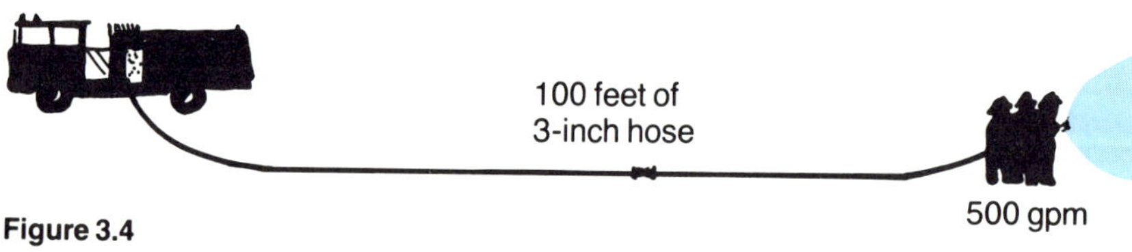

Figure 3.4

$$Q = \frac{gpm}{100}$$

$$Q = \frac{500}{100}$$

$$Q = 5$$

$FL = 2Q^2 + Q$
$FL = 2(5^2) + 5$
$FL = 2(25) + 5$
$FL = 50 + 5$
$FL = 55$ psi in 100 feet of 2½-inch hose

Find conversion factors for 3-inch hose with 2½-inch couplings in Table 3.2.

Either multiply by 0.4 or divide by 2.5.

55(0.4) = 22 psi FL per 100 feet of 3-inch hose with 2½-inch couplings.

$\frac{55}{2.5} =$ 22 psi FL.

Preconnected Hose

On most fire apparatus the total lengths of preconnected booster hose and 1½-, 1¾- and 2-inch hose generally remain the same, although an extra section or two of hose may be added to the lines for wider maneuverability.

Reference to Table 3.3 will enable the pump operator to determine in advance the proper pump discharge pressure to be used when supplying booster lines.

Fire departments are now starting to use 1¾-inch hose with 1½-inch couplings. This new attack line can increase the waterflow from 30 to 50 percent while the pump pressure remains the same.

In many departments 2-inch hose is also being coupled with 1½-inch couplings to facilitate the use of the hose with existing equipment systems. There is no standard friction loss factors for this combination. A department using these sizes of hose should

TABLE 3.3
Friction Loss by Gallons Per Minute Flow
(100 feet of rubber hose)

Diameter	Gallons Per Minute													
(inches)	10	15	20	25	30	35	40	45	50	55	60	65	70	75
¾	14	29	50	Not practical above 20 gpm										
1	4	7	12	19	26	35	44	Not practical above 40 gpm						

run preliminary tests, using flow meters or in-line gauges, to establish the friction loss coefficient for their particular hose, and use this factor when training personnel.

The use of 1¾- and 2-inch hose is meant to deliver more water with the same pump pressure and with as little additional weight as possible, to increase the firefighter's fire fighting efficiency.

Some fog nozzles designed for use with booster lines have been unfairly criticized for their performance when the real fault has been the size of the booster hose. Reference to Table 3.3, again, will indicate that 20 gpm is just about the maximum practical capacity of 100 feet of ¾-inch hose with the pump discharge pressure at 152 psi. If the same nozzle were placed on 200 feet of ¾-inch hose the pump discharge pressure would have to be increased to about 300 psi to obtain the same nozzle performance. The friction loss in ¾-inch booster hose is approximately four times that of 1-inch booster hose with the same quantity of water flowing.

Table 3.4 shows friction loss in fire department hose of good quality, comparing tests run in 1925 with more recent tests. Space is provided for the firefighter to enter the friction loss of the 1½-inch hose used by the local department (there may be some surprising differences).

TABLE 3.4
Friction Loss in 1½-Inch Rubber-Lined Hose
(psi per 100 feet)

Flow (U.S. gpm)	Friction Loss: 1925 Tests	Friction Loss: 1952-58 Tests (avg., 19 samples)	Local Dept. Tests	Flow (L/min) per m
10	0.5	0.3		
20	1.7	1.2		
30	3.6	2.5		
40	6.1	4.3		
50	9.2	7.1		
60	13.0	9.2		
70	17.3	12.3		
80	22.0	15.6		
90	27.3	19.5		
100	33.0	25.5		
120	47.0	33.0		
150	70.0	50.0		

NOTE: The wide variances easily seen in this table suggest that a department should test its hose to find the actual friction loss.

In 1979 the Nebraska State Fire School conducted friction loss experiments with 1½- and 1¾-inch fire hose coupled with 1½-inch couplings. The results are reported here for the purpose of general information. They are not to be taken as hard-and-fast, because of the numerous variables such as type of hose, expansion and interior roughness.

TABLE 3.5
Friction Loss Comparisons
1½- and 1¾-Inch Hose with 1½-Inch Couplings
(psi per 100 feet)

Flow (U.S. gpm)	Friction Loss 1½	1¾	2½
30	3.6	1.15	
60	13.0	5.00	
75	19.0	7.00	
100	33.0	12.00	3.0
125	51.0	18.00	4.5
150	67.0	26.00	6.0
175	95.0	34.00	8.0
200	123.0	45.00	10.0
250		70.00	15.0
300		95.00	21.0

NOTE: These are approximations, not to be used as sure figures. Each department should test its own hose for friction loss.

Source: Nebraska State Fire School.

REQUIRED PUMP DISCHARGE PRESSURE

To determine the pump discharge pressure necessary to produce a good fire stream the firefighter must know the length of the hoseline and the flow from the nozzle. The pump discharge pressure *(PDP)* is equal to the desired nozzle pressure *(NP)* plus the total friction loss *(FL)* in the hose. Therefore:

$$PDP = NP + FL$$

Example 12

What is the required pump discharge pressure when supplying 200 feet of 2½-inch hose equipped with a 200 gpm fog nozzle operating at 100 psi nozzle pressure?

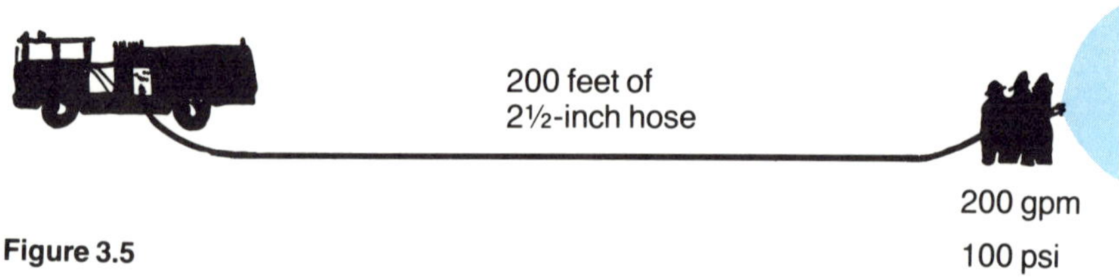

Figure 3.5

$$Q = \frac{gpm}{100}$$

$$Q = \frac{200}{100}$$

$$Q = 2$$

$$FL = 2Q^2 + Q$$
$$FL = 2(2^2) + 2$$
$$FL = 2(4) + 2$$
$$FL = 8 + 2$$
$FL = 10$ psi per 100 feet of 2½-inch hose

FL in 200 feet = 20

$$PDP = NP + FL$$
$$PDP = 100 + 20$$
$PDP = 120$ psi

Multiple Hoselines of Equal Lengths

When more than one hoseline is supplied from the same pumper the hoselines are usually somewhere near the same length. If the nozzles are the same the flow is equal and the friction loss to be overcome by the pumper will be the same. The speed of the pump must be increased to deliver additional water, but the pump discharge pressure will remain the same. When the lines are the same length, calculations for only one line are necessary — the other lines will have approximately the same pressure, as shown in Figure 3.6.

Example 13

Three 2½-inch hoselines, each 400 feet long, are laid with 1-inch nozzle tips. They have a flow pressure of 50 psi each. What is the pump discharge pressure?

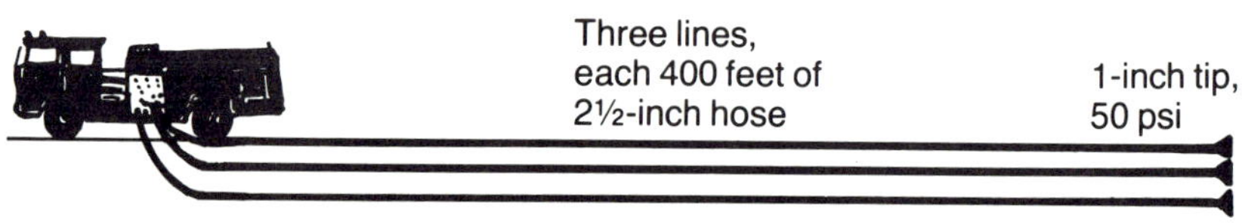

Figure 3.6

$$gpm = 29.7d^2\sqrt{P}$$
$$gpm = 29.7(1^2)(\sqrt{50})$$
$$gpm = 29.7(7.0711)$$
$$gpm = 210 \text{ (round off to 200)}$$

$$Q = \frac{gpm}{100}$$
$$Q = \frac{200}{100}$$
$$Q = 2$$

$$FL = 2Q^2 + Q$$
$$FL = 2(2^2) + 2$$
$$FL = 2(4) + 2$$
$$FL = 8+2$$
$$FL = 10 \text{ psi per 100 feet of } 2\tfrac{1}{2}\text{-inch hose}$$

FL in 400 feet = 4(10) = 40 psi

$$PDP = NP + FL$$
$$PDP = 50 + 40$$
$$PDP = 90$$

The difference between this example and one where only one line is being used is in the *volume at the pump,* not the *pump pressure.* In this example the pumper is supplying three lines at 200 gpm each — a total of 600 gpm.

Exercise 14

What is the pump discharge pressure required when a pumper is supplying four lines of 2½-inch hose, each equipped with a 250 gpm fog nozzle and each 250 feet long?

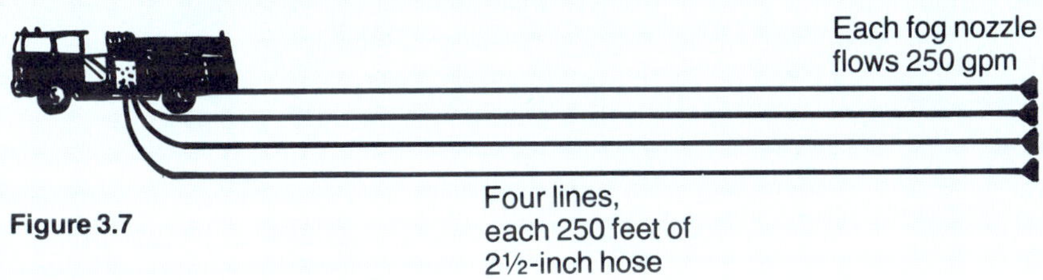

Figure 3.7

Lines of Different Lengths

When there is a considerable difference in the lengths of the hoselines or when the gpm flows are different the same pump discharge pressure for all lines may not produce satisfactory results. In this situation the desired pump discharge pressure must be determined separately for each hoseline and the pumper operated at the highest pressure needed. The pressure supplied to the hoselines requiring lower pressures may be reduced by partially closing the pumper discharge valves until reasonably satisfactory fire streams are being produced. Most new pumpers have individual pressure gauges or flow meters for each discharge outlet to eliminate guesswork when it is necessary to supply hoselines at different outlet discharge pressures.

If one hoseline is considerably longer than other lines a larger capacity nozzle or a larger size tip should be placed on the shortest line when practical. A situation of this nature is illustrated in Figure 3.8. The differences in the flow cause differences in friction in the two lines. These differences tend to equalize the operating pressure if properly laid out. The largest capacity waterflow, which is the longest line, should be used to calculate friction loss.

Example 14

Two 2½-inch hoselines, one 500 feet long, the other 300 feet long. Each has a 250 gpm nozzle operating at 100 psi. What is the required pump discharge pressure?

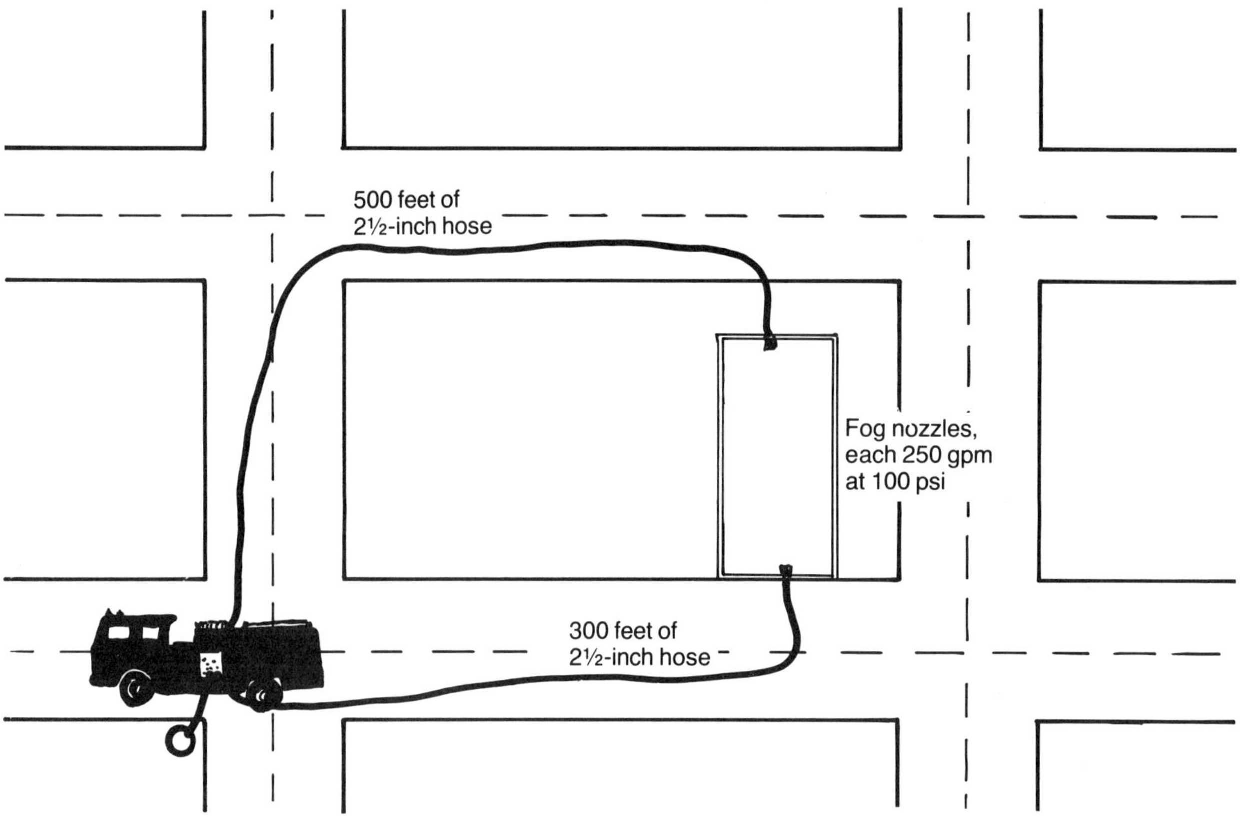

Figure 3.8

$$Q = \frac{gpm}{100}$$

$$Q = \frac{250}{100}$$

$$Q = 2.5$$

$FL = 2Q^2 + Q$
$FL = 2(2.5^2) + 2.5$
$FL = 2(6.25) + 2.5$
$FL = 12.5 + 2.5$
$FL =$ 15 psi per 100 feet of 2½-inch hose

FL in 500 feet = 5(15) = 75 psi
FL in 300 feet = 3(15) = 45 psi

$PDP = NP +$ highest FL
$PDP = 100 + 75$
$PDP =$ 175 psi

Exercise 15

If a pumper is supplying two lines of 2½-inch hose, one 200 feet long and the other 300 feet long, each with a 1-inch nozzle tip, what is the pump discharge pressure when 50 psi is desired at the nozzles?

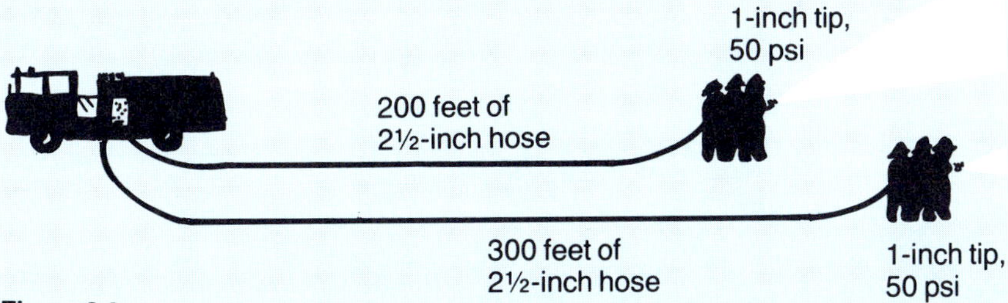

Figure 3.9

NOTE: The pump discharge valve on the short line should be partially closed, which will tend to equalize the nozzle pressures.

Wyed 1½-Inch Lines

A common hose assembly is two 1½-inch fog lines wyed from a 2½-inch supply line. It is the responsibility of the pump operator to supply these two 1½-inch nozzles through the 2½-inch line at 100 psi NP. The total gpm flow from both 1½-inch nozzles at 100 psi must first be established. If the type of nozzle used delivers 100 gpm at 100 psi, then two of these nozzles will deliver 200 gpm. According to the formula $2Q^2 + Q$, the friction loss is 10 psi per 100 feet of 2½-inch hose.

This friction loss, however, is only a part of the total loss. The 1½-inch wye/hose assembly offers additional resistance. The friction loss will vary according to the length of the 1½-inch hose. Usually this assembly is carried as a unit on a fire apparatus and will be the same every time it is used. When a unit of this type is carried a field trial should be made to establish the friction loss in the assembly, as shown in Figure 3.10. This test may be done by placing a gauge at the base of one or both of the nozzles and a gauge between the wye adapter and the 2½-inch hose. The nozzles should be adjusted to a 30-degree pattern because this is an average flow setting. If constant-flow nozzles are used any pattern setting can be used. Pump discharge pressure should be developed to produce 100 psi at the nozzles. The gauge at the wye adapter can then be read to determine the friction loss in the wye/hose assembly, and the pump discharge pressure can be read to check the friction loss per 100 feet of 2½-inch hose.

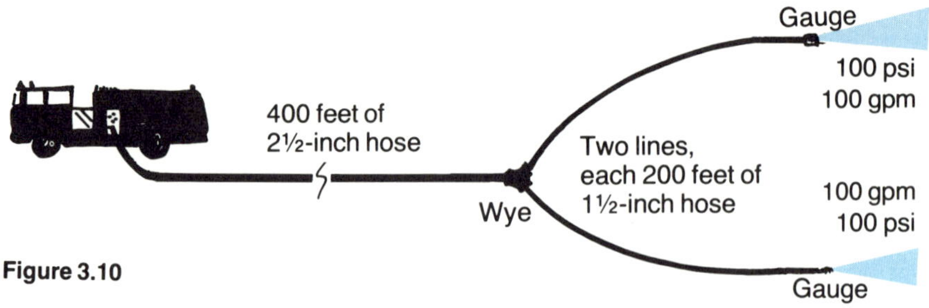

Figure 3.10

$$Q = \frac{gpm}{100}$$

$$Q = \frac{200}{100}$$

$$Q = 2$$

$$FL = 2Q^2 + Q$$
$$FL = 2(2^2) + 2$$
$$FL = 2(4) + 2$$
$$FL = 10 \text{ psi per 100 feet}$$

FL in 400 feet = 40

Friction Loss, 1½-inch

FL in 100 feet of 2½-inch hose at 100 gpm =

$$FL = 2Q^2 + \tfrac{1}{2}Q$$
$$FL = 2(1) + 1$$
$$FL = 3$$

Using conversion factors from Table 3.2:

$$3(13.5) = 40.5$$

or

$$\frac{3}{0.074} = 40.5$$

FL in 200 feet 1½ = 81 psi

Add 10 psi for wye appliance FL:

Assembly $FL = 91$ psi

Total $FL = 40 - 91 = 131$ psi

$$PDP = NP + FL$$
$$PDP = 100 + 131 \text{ psi}$$
$$PDP = 231 \text{ psi}$$

CALCULATE FRICTION LOSS AND PDP FOR LOCAL WYE ASSEMBLY

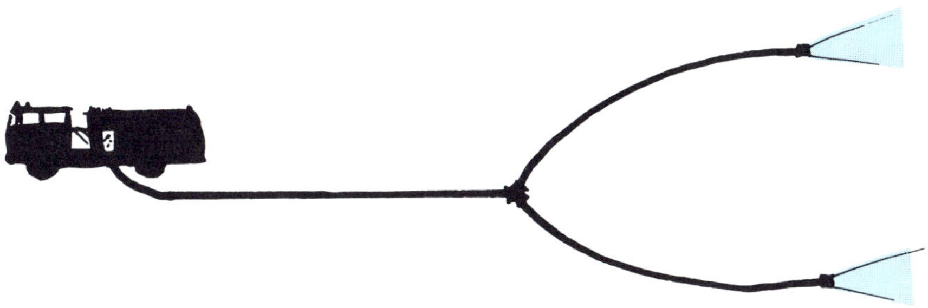

Example 15

Here a field test has shown that there are 25 psi friction loss in the wye assembly (including wye and two 1½-inch lines) when a total of 190 gpm is flowing through two 1½-inch hoselines when each is flowing 95 gpm. The friction loss per 100 feet in the 2½-inch hose is:

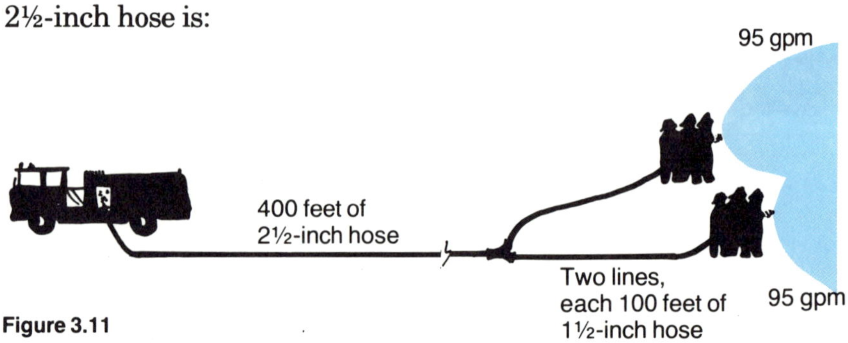

Figure 3.11

$$Q = \frac{gpm}{100}$$

$$Q = \frac{190}{100}$$

$$Q = 1.9$$

$$FL = 2Q^2 + Q$$
$$FL = 2(1.9^2) + 1.9$$
$$FL = 2(3.61) + 1.9$$
$$FL = 7.2 + 1.9$$
$FL = 9$ psi per 100 feet of 2½-inch hose

Once the friction losses have been established for a particular assembly these losses will be the same for every fireground activity if the assembly is not changed. Therefore, the total pump pressure required for the illustrated assembly is:

100 psi	For standard nozzle pressure
25 psi	Loss in the assembly
36 psi	Loss in 400 feet of 2½-inch hose
161 psi	Total pump pressure required

Exercise 16

What is the total friction loss of the layout shown in the drawing below?

Figure 3.12

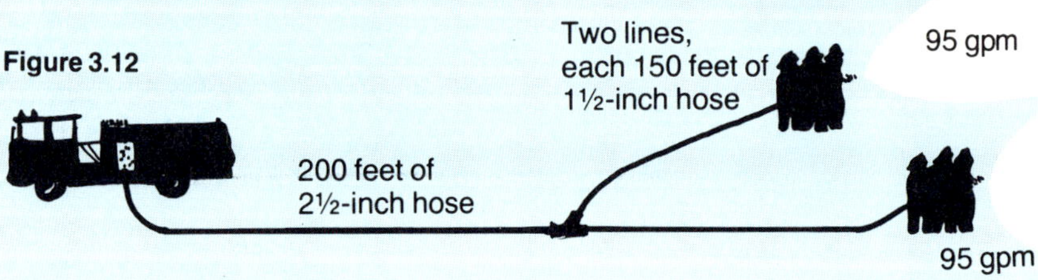

Exercise 17

What pump discharge pressure would be required to supply the layout in the figure below with 100 psi nozzle pressure?

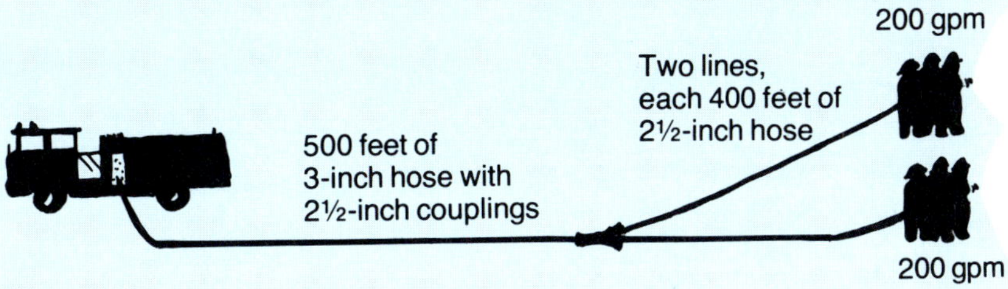

Figure 3.13

Exercise 18

What pump discharge pressure would be required to supply the following layout with 100 psi nozzle pressure?

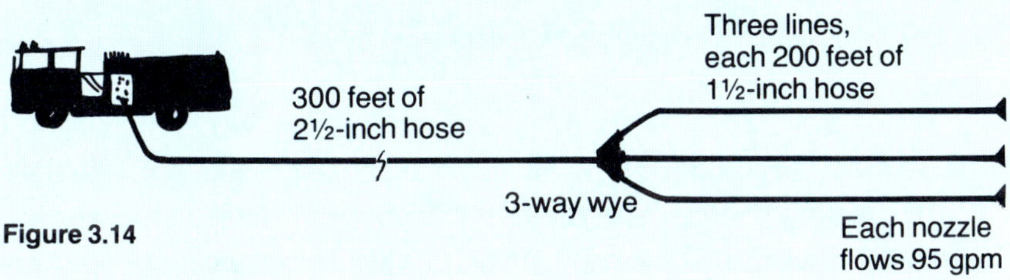

Figure 3.14

Siamesed Hoselines of Equal Length

When the rate of flow is increased through hose additional pressure is required to overcome friction in the hose. When fire streams require great volumes of water, or when hose lays are long, the friction loss in the hose will also be greater. To keep friction loss within reasonable limits firefighters may lay two or more hoselines parallel with each other and siamese them together at a point close to the fire. When two hoselines of equal lengths are siamesed or brought together to supply a master stream the friction loss will be approximately 25 percent of that of a single hoseline at the same nozzle pressure. When three hoselines of equal length are siamesed the friction loss will be approximately 10 percent of that of a single line if equal nozzle pressure is maintained.

Multiple lines of hose laid parallel and siamesed into one line do not present an exceptionally difficult friction loss problem.

Example 16

What is the total friction loss in the following layout?

First, find the friction loss in the single line.

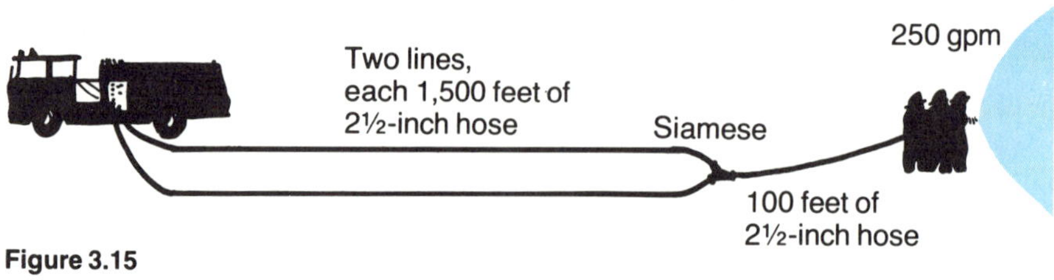

Figure 3.15

$$Q = \frac{gpm}{100} \qquad Q = \frac{250}{100} \qquad Q = 2.5$$

$$FL = 2Q^2 + Q$$
$$FL = 2(2.5^2) + 2.5$$
$$FL = 2(6.25) + 2.5$$
$$FL = 12.5 + 2.5$$
$FL = 15$ psi per 100 feet of 2½-inch hose

There are three ways by which the friction loss in the siamesed hoselines can be determined. Two of these methods make use of the table of conversion factors (Table 3.2). The table shows a division factor of 3.6 and a multiplication factor of 0.28 for converting the friction loss from a single line of two 2½-inch siamesed lines.

$15(0.28) = 4.2$ psi

FL per 100 feet of the siamesed lines

OR

$\frac{15}{3.6} = 4.17$ psi

FL per 100 feet of the siamesed lines

A third method involves a consideration of the flow of water through each siamesed hoseline. When two hoselines of equal lengths are siamesed together only half of the total flow will go through each hoseline. To find the friction loss in one of these lines, use the total flow and divide by 2. Therefore, to find Q, the figure 125 is used.

$$Q = \frac{gpm}{100} \qquad Q = \frac{125}{100} \qquad Q = 1.25$$

$$FL = 2Q^2 + Q$$
$$FL = 2(1.25^2) + 1.25$$
$$FL = 2(1.56) + 1.25$$
$$FL = 3.1 + 1.25$$
$FL = 4.3$ psi per 100 feet of 2½-inch hose

FL in 1,500 feet = 15(4.3) = 65 psi

A siamese device itself also offers resistance to the flow of water, which increases the friction loss. The loss for each siamese device used by a fire department should be found by field testing. Because the friction loss varies from appliance to appliance an average friction loss of 10 psi is used in the examples that follow.

TABLE 3.6
Three Methods of Finding Friction Loss in Siamesed Lines
(Figures from Example 16)

Friction Loss Figures	First Method Divide by 3.6	Second Method Multiply by 0.28	Third Method Use ½ total flow
Loss in siamese	10.0 psi	10.0 psi	10.0 psi
Loss in siamesed lines	62.5 psi	63.0 psi	65.0 psi
Loss in single line	15.0 psi	15.0 psi	15.0 psi
Total loss	87.5 psi	88.0 psi	90.0 psi

The value of using more than one supply line to supply a long layout is very well illustrated in Example 15. If only a single line of 2½-inch hose were used the total friction loss in the 1,500-foot lay would be 15(15), or 225 psi. This friction loss plus the required nozzle pressure may be beyond the pressure at which the hose is subjected for normal fire department tests.

Example 17

Find the pump discharge pressure for the layout below.

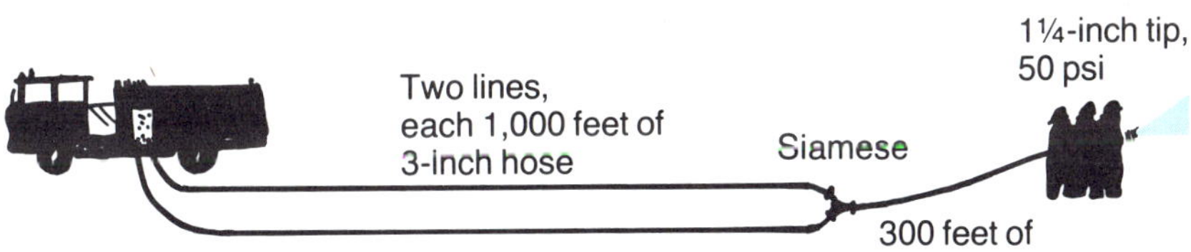

Figure 3.16

$$gpm = 29.7d^2\sqrt{P}$$
$$gpm = 29.7(1.25^2)\sqrt{50}$$
$$gpm = 29.7(1.56)(7)$$
$$gpm = 46.33(7)$$
$$gpm = 324.31$$
$$gpm = 324$$

$$FL = 2Q^2 + Q$$
$$FL = 2(3^2) + 3$$
$$FL = 2(9) + 3$$
$$FL = 18 + 3$$
$FL = 21$ psi per 100 feet of of 2½-inch hose

$$Q = \frac{gpm}{100}$$
$$Q = \frac{324}{100}$$
$$Q = 3.24$$
$$Q = 3$$

FL in 300 feet 2½-inch hose = 3(21) = 63 psi

Conversion factors from Table 3.2 for one 2½-inch hoseline and one 3-inch hoseline: multiply by 0.164 or divide by 6.1.

$$21(0.164) = 3.44 \qquad \frac{21}{6.1} = 3.44$$

FL in 1,000 feet of siamesed 2½-inch and 3-inch hose = 10(3.44) = 34.4 psi.

Total Friction Loss

64.0 psi	FL, single 2½-inch line
34.4 psi	FL, siamesed lines
10.0 psi	FL, siamese appliance (average)
107.4 psi	Total
or	
107 (rounded figure)	

Pump Discharge Pressure

$$PDP = NP + FL$$
$$PDP = 50 + 107$$
$$PDP = 157 \text{ psi}$$

Siamesed Lines of Unequal Length

When lines of unequal length are siamesed, the friction loss is computed by first adding the length of all the lines, then dividing the total length by the number of lines. This average is used as the length for which to find the friction loss.

400 + 500 = 900. 900 divided by 2 = 450.

Example 18

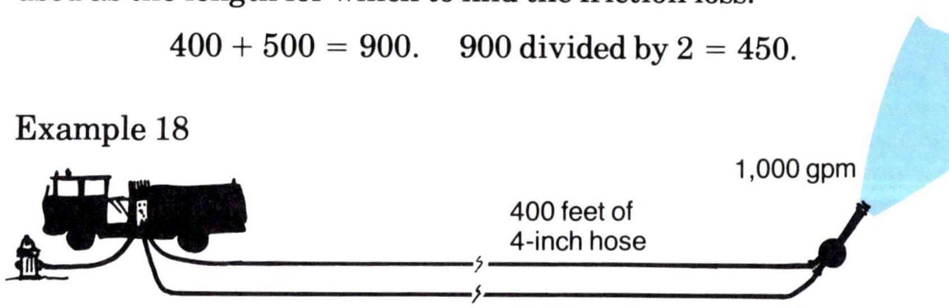

Figure 3.17

$$Q = \frac{gpm}{100} \qquad FL = 2Q^2 + Q$$
$$Q = \frac{1{,}000}{100} \qquad FL = 2(10^2) + 10$$
$$Q = 10 \qquad FL = 200 + 10$$

FL = 210 psi per 100 feet of 2½-inch hose

Using conversion factor from Table 3.2, FL in 100 feet of 4-inch hose is 210(0.09) = 18.9. The FL for 450 feet of 3-inch hose is then 4.5(18.9) = 85.05 psi.

Elevated Lines

If the nozzle is to be elevated the factor of elevation must be taken into account to determine the necessary pump discharge pressure. Using the figures from Example 12, assume the nozzle is to be elevated 50 feet above the pump.

$$EL = 0.5 \text{ psi/ft}$$

The nozzle is to be elevated 50 feet, so the pressure "loss" is

$$0.5(50) = 25 \text{ psi}$$

The pump discharge pressure can be calculated:

$$PDP = NP + FL + EL$$

$$PDP = 100 + 20 + 25$$

$$PDP = 120 + 25$$

$$PDP = 145$$

The use of elevated lines is not limited to advancing handlines up stairs or ladders. Ladder pipes, elevated platforms, telescoping and articulated booms (which are, in effect, modern water towers) are common and need special consideration when friction loss and required pump discharge pressure are determined. The surest method of determining the friction loss of any elevating apparatus is to run tests with each one used by a fire department. Friction loss varies not only from manufacturer to manufacturer, but also in different apparatus by the same manufacturer because the various manifolds and piping configurations (elbows, knuckles, joints) will affect the friction loss.

TABLE 3.7
Friction Loss, Elevating Fire Apparatus Pipe
(psi per 100 feet)

Flow (U.S. gpm)	Pipe Diameter (inches) 4	5	6
500	5.6	1.8	0.7
600	8.7	2.6	1.0
700	10.8	3.4	1.4
800	14.1	4.4	1.7
900	17.7	5.6	2.2
1,000	21.8	6.8	2.7

Source: Warren E. Isman, *Fire Service Pumps and Hydraulics.*

There are no general rules of thumb for friction loss, pump discharge pressure and elevation/angle/attitude/gpm ratios. The maximum allowable nozzle reaction can vary: 400 pounds for a ladder pipe with the nozzle at the tip, 500 pounds for telescoping booms, etc., depending on the manufacturer. All of these items must be found by testing and consultation with the manufacturer. The example below is presented to show how PDP is determined for a common layout. Gpms ranging from 750 to 1,000 are common. The friction-loss figures are not to be taken as absolute figures that would apply to any similar apparatus.

Example 19

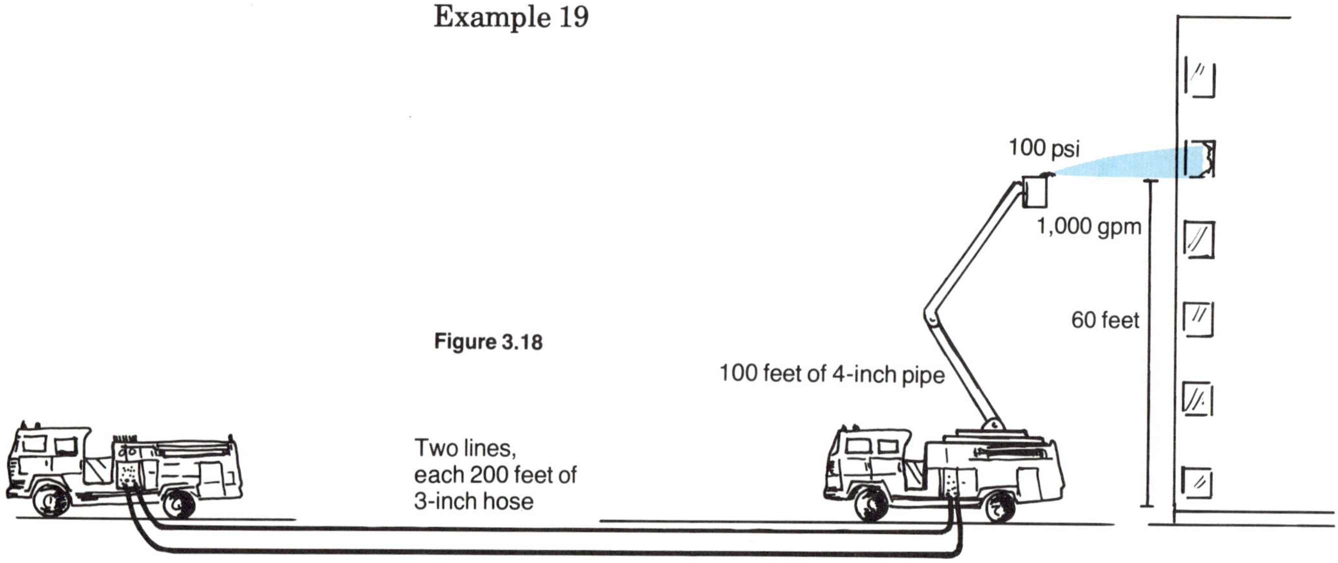

Figure 3.18

Friction Loss, 3-inch hose: Divide the flow by 2: each line carries 500 gpm.

$$Q = \frac{gpm}{100}$$

$$Q = \frac{500}{100}$$

$$Q = 5$$

Manifold FL = 10 psi
Apparatus Piping FL = 5 psi
Elevation Loss = 60(0.5) = 30 psi
Pipe to Platform FL = 21.7 (from Table 3.7)
Nozzle pressure = 100 psi

$$PDP = NP + FL + EL$$
$$PDP = 100+42+10+5+30+21.7$$
$$PDP = 208.7$$

$$FL = (2Q^2 + Q)CF$$
$$FL = (2[5^2] + 5)0.385$$
$$FL = (2[25] + 5)0.385$$
$$FL = 55(0.385)$$
FL = 21 psi per 100 feet of 3-inch hose

FL in 200 feet = 42 psi

Master Streams

Since the volume flowing per minute is the factor that determines friction loss in hose, one has only to know the capacity of each master stream nozzle used by the local department at the standard nozzle pressure. The exact volume per minute is not important. Only the approximate flow need be remembered. The comparative flows for master solid streams at 80 psi nozzle pressure are listed in Table 3.8.

FLOWS THAT WARRANT MASTER STREAMS

Since 400 gpm (1,500 L/min) produces an excessive friction loss in one line of 2½-inch (65 mm) hose (36 psi FL per 100 feet or 250 kPa per 30 m), the flow should be divided through two

TABLE 3.8
Approximate Flows for Master Streams at 80 PSI

Nozzle Tip Size (in inches)	Actual Flow (U.S. gpm)	Approximate Flow
1¼	413	400
1⅜	500	500
1½	596	600
1⅝	700	700
1¾	813	800
2	1,063	1,000

hoselines so that one-half the volume (200 gpm or 750 L/min) is flowing through each line. This flow is comparable to a 1-inch tip at 50 psi (350 kPa) nozzle pressure that also has a flow of 200 gpm (750 L/min). The friction loss, therefore, is the same in each case: 10 psi (70 kPa) per 100 feet (30 m). The same principle is true with a 1½-inch (38 mm) tip delivering 600 gpm (2,250 L/min). The 600 gpm can be divided into three 2½-inch hoselines and thus permit 200 gpm to flow through each line. As a result, the friction loss is still 10 psi per 100 feet of hose.

Table 3.9 illustrates how various flows can be divided into siamesed hoselines to reduce friction loss. An individual should study those flows that pertain to the nozzles used by that individual's department. Such a study will establish the number of hoselines required to produce master streams with a minimum friction loss. When a firefighter is using the chart to determine pump pressure an allowance should be added for friction loss in the master stream device. Allowance for elevation should also be added when a ladder pipe is used. Allowance for elevation is exactly the same as for handlines, 5 psi (35 kPa) per story or 0.5 psi per foot (3 kPa per 0.3 m) of elevation above ground level.

When it is necessary to use other nozzle pressures the following rule of thumb may be used with reasonable accuracy to determine the friction loss per 100 feet of hose. Again it must be remembered that extreme accuracy is not required for fireground calculations and that reasonably good fire streams will be obtained even though the nozzle pressure may be a few psi or kPa high or low.

For a single line of 3-inch (75 mm) hose the friction loss will be about 1½ times as great as for an equal length of two lines of 2½-inch (65 mm) hose, or only 0.4 of the friction loss in a single line of 2½-inch hose.

For three lines of 2½-inch hose friction loss will be about 0.5 of the friction loss in two lines of 2½-inch hose, or 0.1 of one line.

For four lines of 2½-inch hose the friction loss will be about ¼ of the friction loss in two lines of 2½-inch hose, or 0.06 of one line.

NOTE: The friction loss in each master stream appliance is different. An average figure of 10 psi (70 kPa) is used in the problems presented in this manual. The friction loss in the department's master stream devices should be determined. All of the problems given here can be worked by rule-of-thumb methods where applicable (see "Rules of Thumb," presented later in this chapter).

TABLE 3.9
Hoselines and GPM for Producing Master Streams
(80 psi nozzle pressure)

Number and Sizes of Hoselines		GPM Flow per Hoseline	Approximate Total GPM Flow	Friction Loss per 100 Feet (psi)	Nozzle Tip Size
One 2½-inch	(not practical)	400	400	36	1¼
Two 2½-inch	(desirable)	200	400	10	1¼
One 3-inch	(desirable)	400	400	15	1¼
One 2½-inch	(not practical)	500	500	55	1⅜
Two 2½-inch	(desirable)	250	500	15	1⅜
One 3-inch	(less desirable)	500	500	22	1⅜
Two 2½-inch	(less desirable)	300	600	21	1½
Three 2½-inch	(desirable)	200	600	10	1½
Two 3-inch	(desirable)	300	600	8	1½
Two 2½-inch	(not practical)	350	700	29	1⅝
Three 2½-inch	(desirable)	233	700	13	1⅝
Two 3-inch	(desirable)	350	700	12	1⅝
Three 2½-inch	(less desirable)	270	800	18	1¾
Four 2½-inch	(desirable)	200	800	10	1¾
Two 3-inch	(desirable)	400	800	15	1¾
Four 2½-inch	(desirable)	250	1,000	15	2
Three 3-inch	(desirable)	335	1,000	10	2

Example 20

Find the pump discharge pressure when two lines of 2½-inch hose 300 feet long feed a master stream with a 1¼-inch tip at 80 psi NP.

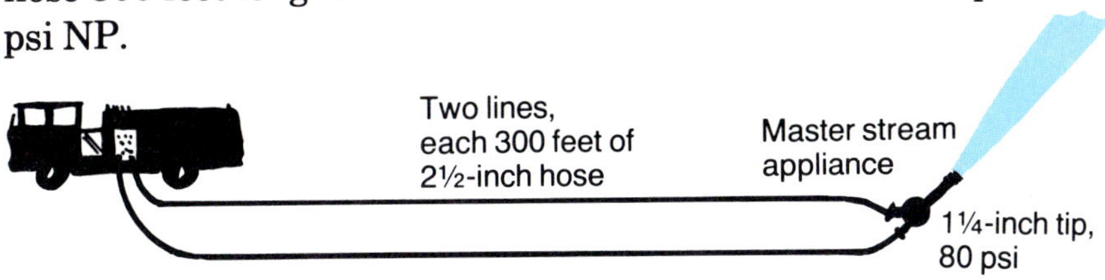

Figure 3.19

The flow of 400 gpm can be divided equally between the two siamesed lines, which will give a flow of 200 gpm. This flow is the same as a 1-inch tip at 50 psi and the FL will also be 10 psi per 100 feet of hose.

FL in 300 feet of 2½-inch hose = 3(10) = 30 psi.

$$PDP = NP + FL + AL$$
$$PDP = 80 + 30 + 10$$
$$PDP = 120 \text{ psi}$$

Example 21

Find the pump discharge pressure in the following problem:

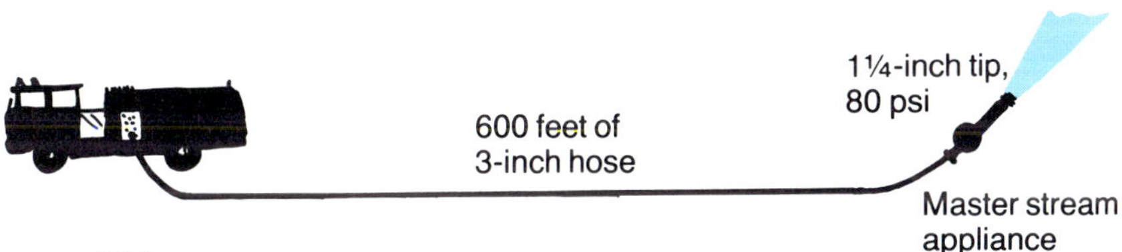

Figure 3.20

$$gpm = 29.7d^2\sqrt{P}$$
$$gpm = 29.7(1.25^2)(\sqrt{80})$$
$$gpm = 29.7(1.265)(8.9443)$$
$$gpm = 37.5(8.9443)$$
$$gpm = 335$$

$$Q = \frac{gpm}{100}$$

$$Q = \frac{335}{100}$$

$$Q = 3.35$$

$$FL = 2Q^2 + Q$$
$$FL = 2(3.35^2) + 3.35$$
$$FL = 2(11.22) + 3.35$$
$$FL = 22.44 + 3.35$$
$$FL = 26 \text{ psi per 100 feet of 2½-inch hose}$$

FL in 600 feet of 2½-inch hose = 6(26) = 156 psi

Conversion factors from Table 3.2

$$\frac{156}{2.6} = 60 \text{ or } 156(0.385) = 60$$

FL in 600 feet of 3-inch hose = 60 psi

$$PDP = NP + FL + AL$$
$$PDP = 80 + 60 + 10$$
$$PDP = 150 \text{ psi}$$

Example 22

A pumper at the base of a hill is feeding a master stream device on the roof of a one-story building at the top of the hill. What pump discharge pressure would be required?

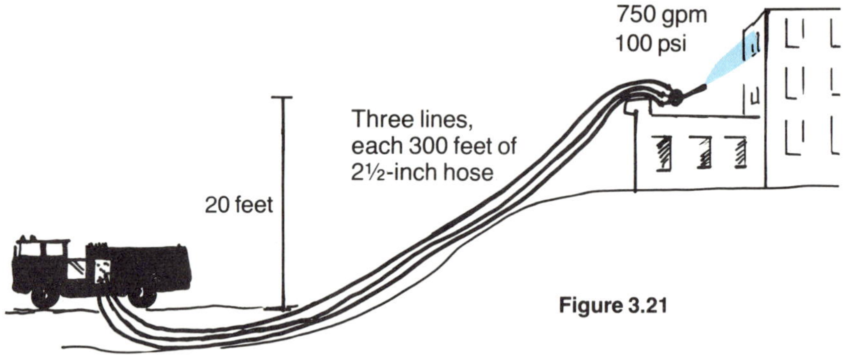

Figure 3.21

The flow of 750 gpm can be divided between the three hoselines, which will give a 250 gpm flow through each line. FL in 2½-inch hose with 250 gpm flow is 15 psi per 100 feet.

FL in 300 feet of 2½-inch hose = 3(15) = 45 psi

Pressure "loss" from elevation calculated as follows:

10 psi	20 feet at 0.5 psi/ft.
5 psi	1 story above ground level
15 psi	Total

$$PDP = NP + FL + AL + EL$$
$$PDP = 100 + 45 + 10 + 15$$
$$PDP = 100 + 70$$
$$PDP = 170 \text{ psi}$$

Example 23

Find the pump discharge pressure when two lines of 2½-inch hose 600 feet long feed a master stream device with a 500 gpm fog nozzle at 100 psi.

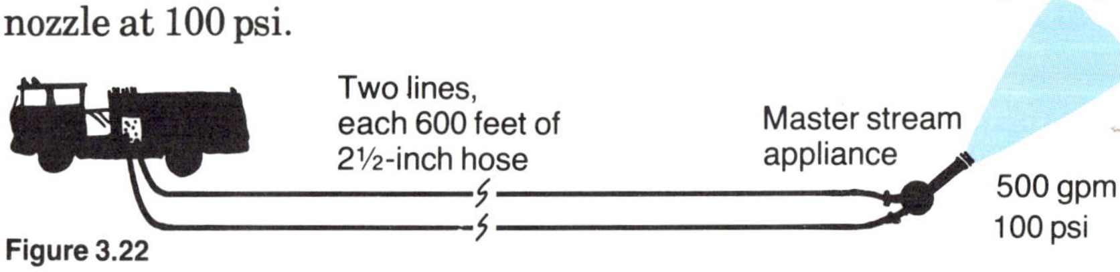

Figure 3.22

The flow of 500 gpm can be divided by 2, which will give 250 gpm through each hoseline. The FL with 250 gpm flowing in 2½-inch hose is 15 psi per 100 feet.

FL in 600 feet of 2½-inch hose = 6(15) = 90 psi

$$PDP = NP + FL + AL$$
$$PDP = 100 + 90 + 10$$
$$PDP = 200 \text{ psi}$$

Example 24

A pumper is supplying three 2½-inch lines 300 feet long. A 750 gpm fog nozzle is being used at 100 psi. What is the pump discharge pressure?

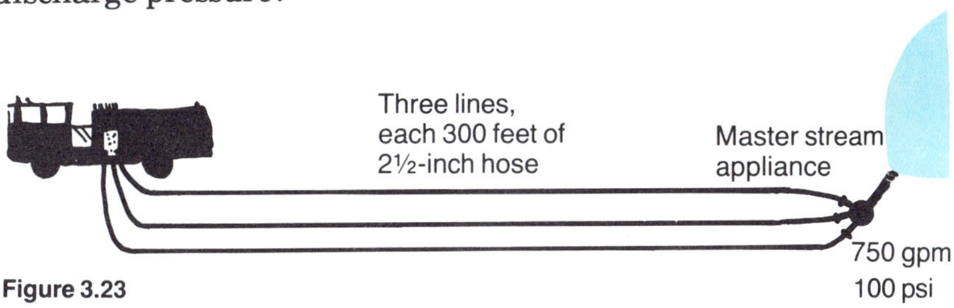

Figure 3.23

First find FL for 2½-inch hose.

The flow of 750 gpm can be divided between the three hoselines, which will give 250 gpm per hoseline. FL in 2½-inch with 250 gpm flowing is 15 psi per 100 feet. FL in 300 feet of 2½-inch hose = 3(15) = 45 psi.

$$PDP = NP + FL + AL$$
$$PDP = 100 + 45 + 10$$
$$PDP = 155 \text{ psi}$$

Example 25

What will be the pump discharge pressure to supply the following layout?

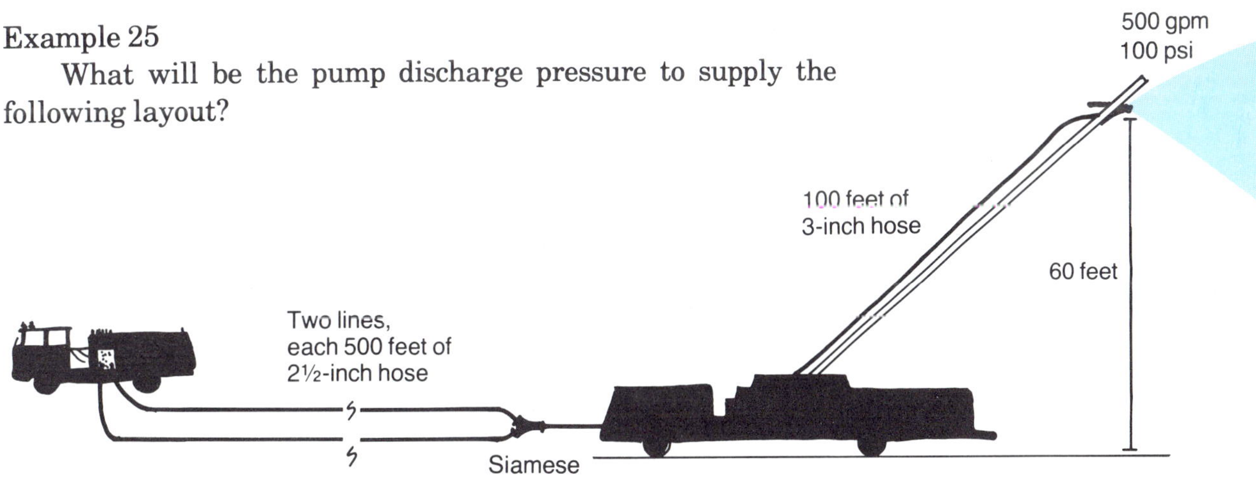

Figure 3.24

The flow of 500 gpm can be divided between the two 2½-inch lines, which will give 250 gpm through each line. FL with 250 gpm flowing is 15 psi per 100 feet. FL for 500 feet of 2½-inch hose = 5(15) = 75 psi.

FL for a single line of 3-inch hose = 5^2 = 25 psi per 100 feet
Loss due to 60 feet elevation = 0.5 psi per foot = 30 psi
FL for siamese: use average figure of 10 psi
FL for ladder pipe: use average figure of 10 psi

Total Loss

30 psi	Elevation "loss"
20 psi	Appliance loss
75 psi	2½-inch hose
25 psi	3-inch hose
150 psi	Total

$$PDP = NP + FL$$
$$PDP = 100 + 150$$
$$PDP = 250 \text{ psi}$$

Friction Loss in Large Diameter Hose

The use of large diameter hose (3- to 6-inch, 75 to 150 mm) has been common for many years in Europe, and has become more popular in the United States. Large volumes of water can flow great distances with relatively low pressure when large diameter hose is used. This allows a pumper to be positioned at the fire, several hundred feet from the source of water supply, which can free a pumper that might otherwise be used in relaying water. Large diameter hose can also be used as the first several lengths of an attack line for supplying smaller handlines and for portable master stream appliances.

TABLE 3.10
Friction Loss, Large Diameter Hose
(psi per 100 feet)

Flow	Hose Diameter (inches)				
(U.S. gpm)	3	3½	4	5	6
200	3.9	1.8	.82	.28	.12
250	5.9	2.7	1.40	.47	.18
300	8.2	3.7	1.90	.66	.24
400	14.1	6.3	3.30	1.10	.42
500	21.2	9.5	5.00	1.70	.62
700	39.5	17.7	9.60	3.30	1.20
800	50.5	22.7	12.40	4.30	1.70
1,000	76.5	34.3	19.20	6.60	2.60
1,100	91.5	41.0	23.00	7.90	3.00
1,200			28.50	9.50	3.50
1,500			41.50	14.50	5.70
2,000			75.00	26.00	12.00

Table 3.10 shows distances and flow rates possible using large diameter hose. Note, for instance, that approximately six times as much water will flow through 100 feet (30 m) of 5-inch hose, with the same friction loss, as will flow through 100 feet of 2½-inch hose.

Table 3.11 shows distances and flow rates possible using large diameter hose. Note, for instance, that approximately six times as much water will flow through 100 feet (30 m) of 5-inch hose, with the same friction loss, as will flow through 100 feet of 2½-inch hose.

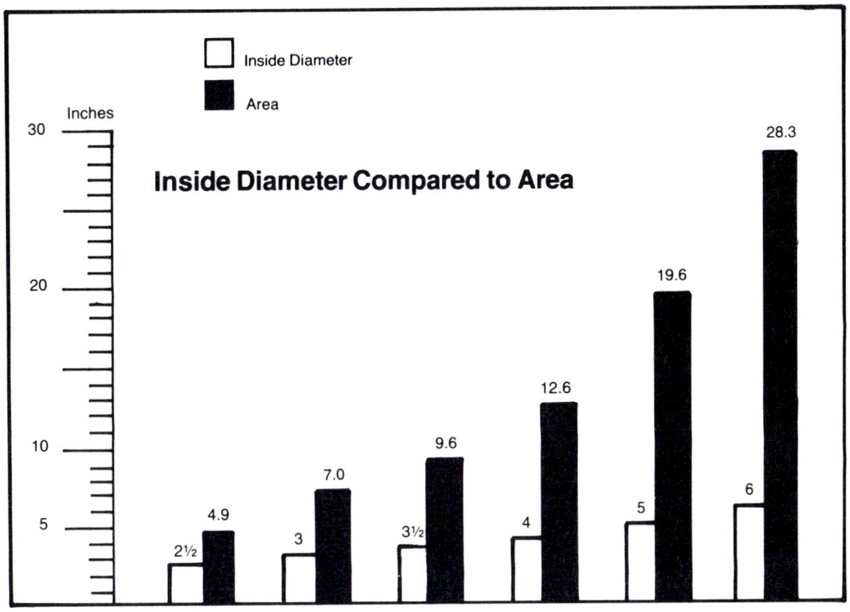

Figure 3.25 A relatively small increase in diameter results in a large increase in area.

TABLE 3.11
Length of Hoseline Possible at 20 PSI Compound
(in feet)

Hose Diameter (inches)	Flow (U.S. gpm)	Hydrant Pressure 40	50	60	70	80	90	100
2½	500	50	50	50	75	100	100	150
	700							
	1,000							
3	500	100	150	200	250	300	350	400
	700	50	75	100	100	150	150	200
	1,000			50	50	50	75	100
3½	500	200	300	400	500	600	700	800
	700	100	150	200	250	300	350	400
	1,000	50	50	100	100	150	200	200
4	500	400	600	800	1,000	1,200	1,400	1,600
	700	200	300	400	500	600	700	800
	1,000	100	150	200	250	300	350	400
5	500	1,150	1,800	2,300	2,900	3,500	4,100	4,700
	700	600	900	1,200	1,500	1,800	2,100	2,400
	1,000	300	450	600	750	900	1,050	1,200
6	500	3,200	4,800	6,400	8,000	9,600	11,200	12,800
	700	1,650	2,400	3,300	4,100	5,000	5,800	6,600
	1,000	750	1,100	1,500	1,900	2,300	2,600	3,000

NET ENGINE PRESSURE

Net engine pressure (also called net pump pressure) takes into account all factors that contribute to the amount of work that the pump must do to produce a fire stream.

At a hydrant the net engine pressure is the difference between the pump discharge pressure and the incoming pressure from the hydrant. If the PDP, for instance, is 150 psi and the intake gauge reads 50 psi, the net engine pressure is 100 psi.

When drafting, the net engine pressure is the sum of the pump discharge pressure, the intake hose friction loss and the intake pressure. A pump expends as much energy to lift water a certain distance as it does to force the same amount of water upward the same distance. In other words, a pump operating at a 10-foot lift and forcing the lifted water 50 feet above the pump is doing the same amount of work as it would to force the water 60 feet upward. (Remember that lift pressure is negative pressure, whereas the pressure of water flowing from a hydrant is positive pressure.)

Example 26

A pump is discharging water at 120 psi. The friction loss in the intake hose is 8 psi. The intake gauge reads 20 inches of mercury (converts to 10 psi).

120 psi	(PDP)
8 psi	(FL)
10 psi	(intake reading)
138 psi	Net engine pressure

COMPUTING AVAILABLE WATER AT A HYDRANT

When a pumper is hooked up to a hydrant and there is no water flowing the reading at the intake gauge is the *static pressure* of the water supply at that point.

When water is being pumped the reading will fall. This decrease, as a percentage, is important to know — it indicates how much more water is available from the hydrant (see Table 3.12).

TABLE 3.12
Additional Water Available at a Hydrant

Percent Decrease of Pressure	Additional Water Available
0 - 10	3 times amount being delivered.
11 - 15	2 times amount being delivered.
16 - 25	Same amount as being delivered.
25+	More water might be available, but not as much as is being delivered.

From Table 3.12 one can see that if one line is being supplied and there is less than a 10 percent drop in static pressure, three more lines like the first one can be supplied.

COMPUTING MAXIMUM LIFT

Lift is measured from the surface of the static source to the centerline of the pump. The height of possible lift is not affected by the angle of the intake hose, but it is affected by the vacuum that the pump can produce and by the atmospheric pressure.

Theoretically, at sea level a pump can lift water 33.8 feet (14.7 psi times 2.3 feet per pound). A perfect vacuum is impossible with a fire pump and there will be friction loss, so the maximum lift would be nearer to 20 to 25 feet.

The height that water can be lifted decreases with altitude by about 1 foot for each 1,000 feet of elevation. The weather also affects drafting, but to a smaller degree.

To determine the height to which water can be drafted, the vacuum reading in inches of mercury at the compound gauge of the pumper is read and multiplied by 0.49. The result is then multiplied by 2.3 because water rises 2.3 feet for each psi removed.

The formula for determining maximum lift is

$$\mathrm{h} = 1.13Hg$$

where h = height of lift (in feet)
Hg = inches of vacuum

TABLE 3.13
Friction Loss Allowances, Intake Hose

Rated Capacity of Pumper (U.S. gpm)	Hose Diameter (inches)	Allowancc (feet): For 10 Feet of Hose	Allowancc (feet): Each Additional 10 Feet
500	4	6	+1
	4½	3½	+½
750	4½	7	+1½
	5	4½	+1
1,000	4½	13	+2
	5	8	+1½
	6	4	+½
1,250	5	12½	+2
	6	6½	+½
1,500	6	9	+1

Source: American Insurance Association

A good rule of thumb is that for each inch of indicated vacuum, water will rise one foot in the intake hose and pump.

TABLE 3.14
Friction Loss, 20 Feet of Intake Hose
(in psi)

Hose Diameter (inches)	Area in Square Inches	Flow (U.S. gpm)				
		500	750	1,000	1,250	1,500
2½	4.909	11.0	24.0	42.0	65.0	93.0
Two 2½	9.818	3.0	6.3	11.0	16.8	24.0
3	7.068	4.4	8.6	16.8	26.0	37.2
4	12.566	1.0	2.1	3.7	5.8	8.3
4½	15.896	0.56	1.2	2.1	3.3	4.7
5	19.635	0.34	0.75	1.3	2.0	2.9
6	28.274		Negligible		Negligible	2.2

Nozzle Reaction

Newton's Third Law of Motion — For every action there is an equal and opposite reaction — is very evident at the fire hose nozzle, where its results can do the most harm to firefighters and to the fire attack strategy. A recoilless nozzle has yet to be invented, so the firefighters at the nozzle take the full force of the nozzle reaction. One of the pump operator's main responsibilities is to be sure that the nozzle reaction is small enough to be handled. The larger the volume, the greater the reaction; the greater the nozzle pressure, the greater the reaction.

The formula for the nozzle reaction of a solid stream nozzle is:

$$NR = 1.5d^2NP$$

where

NR = nozzle reaction (pounds)
d = nozzle diameter (inches)
NP = nozzle pressure (psi)

Example 27

A ⅞-inch tip with a nozzle pressure of 50 psi.

$$NR = 1.5(0.875^2)(50)$$
$$NR = 1.5(0.765)(50)$$
$$NR = 1.14(50)$$
$$NR = 57 \text{ pounds}$$

The nozzle reaction of handlines is always somewhat less than theoretical, but a nozzle directed horizontally from a ladder exerts full theoretical nozzle reaction, which can force the ladder (and any person on it) away from the building.

The maximum nozzle reaction that can be handled by three firefighters at the nozzle is 95 pounds. The maximum recommended nozzle reaction for a ladder pipe mounted on the top fly is 400 pounds. (Consult with manufacturer of locally used apparatus.)

Nozzle reaction is less severe with fog nozzles because the water leaves the nozzle at angles, which tends to cancel out some of the reactionary forces. The reaction and flow will be greatest at a straight stream setting.

The formula for fog nozzle reaction is:

$$NR = 0.0505Q\sqrt{NP}$$

where

NR	=	nozzle reaction (pounds)
Q	=	flow (gpm)
NP	=	nozzle pressure (psi) at *base* of the nozzle

Example 28

A 1½-inch fog nozzle flowing at 95 gpm with a nozzle pressure of 100 psi.

$$NR = 0.0505(95)(\sqrt{100})$$
$$NR = 0.0505(950)$$
$$NR = 47.98 \text{ pounds}$$

A rule-of-thumb calculation for fog nozzle reaction is:

$$NR = \tfrac{1}{2}gpm$$

To safely and successfully control this reaction the firefighters at the nozzle should use certain techniques for handling the nozzle. These techniques are presented in IFSTA 103, **Fire Hose Practices.**

GPM FLOWS FROM FOG NOZZLES

The discharge of any fixed-orifice nozzle varies directly at the square root of the nozzle pressure. The volume of water being discharged from a nozzle at any pressure can be determined if the flow is known at a given nozzle pressure. This ratio may be stated as:

$$\frac{gpm_1}{\sqrt{NP_1}} = \frac{gpm_2}{\sqrt{NP_2}}$$

$$gpm_2 = \frac{(\sqrt{NP_2})(gpm_1)}{\sqrt{NP_1}}$$

where

gpm_1 = Rated discharge
NP_1 = Rated nozzle pressure
gpm_2 = Operating discharge
NP_2 = Operating nozzle pressure

Since most fog nozzle discharges are rated at 100 psi, their discharge at other pressures can be determined by using the above formula written as follows:

$$gpm_2 = 0.1(\sqrt{NP_2})(gpm_1)$$

Example 29

A fog nozzle is rated at 250 gpm discharge at 100 psi nozzle pressure. What is the discharge from this nozzle when operating at a nozzle pressure of 81 psi?

$$gpm_2 = 0.1(\sqrt{NP_2})(gpm_1)$$
$$gpm_2 = 0.1(\sqrt{81})(250)$$
$$gpm_2 = 0.1(9)(250)$$
$$gpm_2 = 0.9(250)$$
$$gpm_2 = 225$$

Exercise 19

A given fog nozzle is rated 30 gpm. What is the discharge from this nozzle when operating at a nozzle pressure of 225 psi?

RULES OF THUMB FOR FRICTION LOSS

Fireground activities often prohibit the use of formulas because of existing conditions and the lack of time, so rules of thumb can be used when it is necessary to calculate quickly. There are several rules of thumb used for quick calculations, and the most commonly used are presented here. The order by which these rules are presented in no way signifies their value or importance. Their adaptation must be made locally and to the best advantage of all concerned. Metric equivalents are not given in this section as no metric rules of thumb were known at the time of printing.

Condensed Q Formula for 3-Inch Hose

Some fire departments use 3-inch hose with 2½-inch couplings to supply master streams. Three-inch hose is especially beneficial when supplying ladder pipes because of the reduction of friction loss. The friction loss per 100 feet of 3-inch hose with 2½-inch couplings may be determined by the formula $FL = Q^2$ when the flow is less than 500 gpm. The symbol Q must be stated in the number of hundreds of gallons per minute, or gpm divided by 100. This method and formula is not as accurate as using the conversion factor from Table 3.2 but it is more rapid and is sufficiently accurate for fireground operations. For practical purposes, 3-inch hose with either 3-inch or 2½-inch couplings can be treated identically.

Example 30

What will be the friction loss per 100 feet of 3-inch hose when there are 200 gpm flowing?

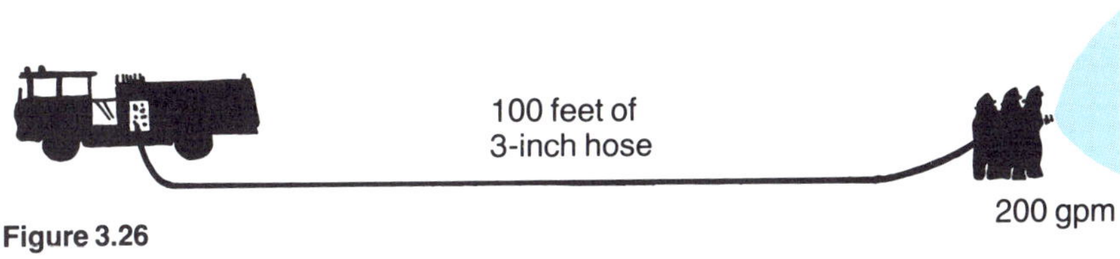

Figure 3.26

$$Q = \frac{gpm}{100} \qquad FL = Q^2$$

$$Q = \frac{200}{100} \qquad FL = 2^2$$

$$Q = 2 \qquad \text{FL} = 4 \text{ psi per 100 feet of 3-inch hose}$$

Example 31

What will be the friction loss per 100 feet of 3-inch hose when there are 500 gpm flowing?

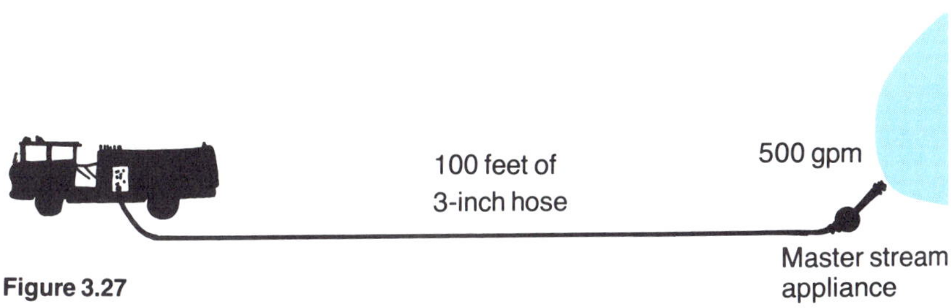

Figure 3.27

$$Q = \frac{gpm}{100} \qquad FL = Q^2$$

$$Q = \frac{500}{100} \qquad FL = 5^2$$

$$Q = 5 \qquad \text{FL} = 25 \text{ psi per 100 feet of 3-inch hose}$$

Example 32

What will be the pump discharge pressure required to supply a 500 gpm fog nozzle on 600 feet of 3-inch hose with 2½-inch couplings?

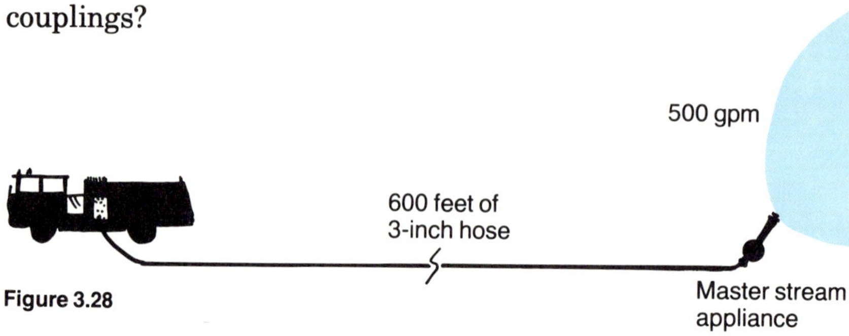

Figure 3.28

$$Q = \frac{gpm}{100}$$

$$Q = \frac{500}{100}$$

$$Q = 5$$

$$FL = Q^2$$
$$FL = 5^2$$
$$\text{FL} = 25 \text{ psi per 100 feet of 3-inch hose}$$

$$\text{FL in 600 feet} = 6(25) = 150 \text{ psi}$$

$$PDP = NP + FL$$
$$PDP = 100 + 150$$
$$PDP = 250$$

GPM Flowing

The following rule-of-thumb method permits friction loss to be calculated from the gpm flow and is applicable to any solid or fog stream. Another desirable feature of this method is that it can be used for sizes of hose other than 2½-inch and can apply to smaller sizes of hose if sufficient study is given to its use.

In Table 3.15 note that a line separates the flow in gpm between 150 and 170. This separation is necessary because the rule of thumb does not apply to 2½-inch hose with flows below 160 gpm. Flows less than 160 gpm are also seldom encountered through 2½-inch hose. A firefighter needs to know only the flow in gpm from a nozzle at a specified pressure. Then, by subtracting ten from the first two numbers of the gpm flow a sufficiently accurate friction-loss figure per 100 feet of 2½-inch hose can be obtained. A study of Table 3.15 further reveals that friction loss in 2½-inch hose increases one psi for every 10 gpm increase in flow.

In Table 3.16 the rule of thumb is applied to 1½-inch hose. Table 3.16 is divided into three distinct sections. The upper section represents flows from 50 to 75 gpm. This section is designed to show that friction loss in 1½-inch hose is the same as it is in 2½-inch hose with four times as much water flowing. For example: One desires to know the friction loss in 1½-inch hose when 70 gpm are flowing; therefore, 4(70) = 280. Using Table 3.15, subtract 10 from 28. The result is 18 psi friction loss per 100 feet in 1½-inch hose.

TABLE 3.15

GPM Flowing 2½″ Hose	FL per 100 ft. 2½″ Hose	Note
100	3	
110 - 120	4	
130 - 140	5	
150	6	
160 - 170	7	Subtract "10" from the first two numbers of gpm
180	8	
190	9	
200	10	
210	11	
220	12	
230	13	
240	14	
250	15	
260	16	
270	17	
280	18	
290	19	
300	20	

TABLE 3.16

GPM Flowing 1½″ Hose	GPM Flowing 2½″ Hose	FL per 100 ft. 1½″ or 2½″ Hose	Note
50	200	10	Same as 4 times as much water flowing in 2½-inch; For increase of 20 gpm flowing FL increases 2 psi
55	220	12	
60	240	14	
65	260	16	
70	280	18	
75	300	20	
80	320	23	For increase of 20 gpm flowing FL increases 3 psi
85	340	26	
90	360	29	
95	380	32	
100	400	35	
105	420	39	For increase of 20 gpm flowing FL increases 4 psi
110	440	43	
115	460	47	
120	480	51	
125	500	55	

The middle section of Table 3.16 represents flows from 80 to 100 gpm. Four times the flow through 1½-inch hose is shown in the 2½-inch column. The friction loss in 1½-inch hose, however, simply increases by three psi for every 20 gpm increase in flow through 2½-inch hose.

The lower section of Table 3.16 represents flows from 105 to 125 gpm. As in the top two sections, four times the flow through 1½-inch hose is shown for 2½-inch hose. The friction loss for these pressures increases four psi for every 20 gpm increase in flow through 2½-inch hose.

When consideration is given to friction loss in booster hose it should be remembered that the friction loss in 1-inch booster hose is approximately seven times that of 1½-inch hose and that the friction loss in ¾-inch booster hose is approximately four times that of 1-inch booster hose with the same quantity of water flowing. Rule-of-thumb formulas for determining friction loss in booster hose are: 1-inch FL = gpm − 4; ¾-inch FL = (gpm − 4)(4).

Counting Fingers

Fire stream calculations on the fireground require methods that are easily understood and applicable to mnemonic use. The following method of using one's hand and fingers applies only to 2½-inch hose for flows from 100 to 500 gpm.

Starting with the thumb of the left hand, as illustrated in Figure 3.29,, each finger is numbered at the tip in terms of hundreds of gallons per minute. Returning to the thumb, and again moving from left to right, the base of each finger is given a successive odd number, beginning with three. The friction loss for 100 feet of 2½-inch hose at a desired flow is determined by selecting the finger to which the desired flow has been assigned, and multiplying the number at the tip of the finger by the number at the base of the finger. Thus, the friction loss for a flow of 500 gpm can be found by using the numbers assigned to the little finger, or 5(11) = 55 psi friction loss per 100 feet of 2½-inch hose. Likewise, the friction loss for a flow of 200 gpm is found using the numbers assigned to the index finger. Thus, 2(5) = 10 psi friction loss per 100 feet of 2½-inch hose.

Since nozzle capacities vary in gpm, the nearest half-hundred can be used with slight variations. The numbers 1.5, 2.5, 3.5 and 4.5 can be used for flows of 150, 250, 350 and 450 gpm, respectively. These half-hundred figures can be assigned to the spaces between the fingers. Then by using the even numbers 4, 6, 8 and 10 to couple with the half-hundred fingers, as shown in Figure 3.30, sufficiently accurate results can be obtained for the friction loss per 100 feet of 2½-inch hose. The application for both systems to one hand is illustrated in Figure 3.31.

The counting method can also be used for 3-inch hose by numbering the fingers as shown in Figure 3.32.

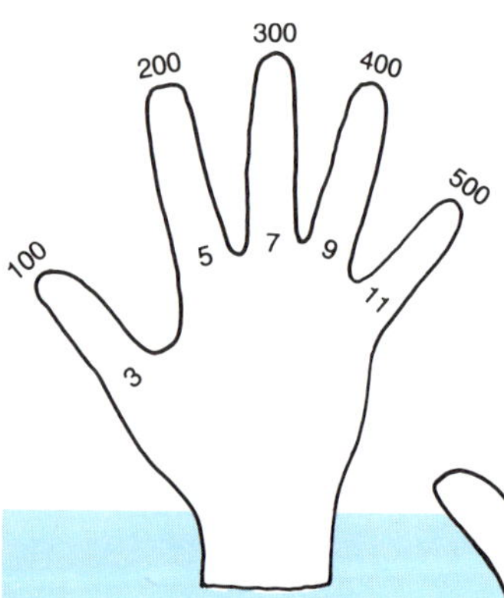

Figure 3.29

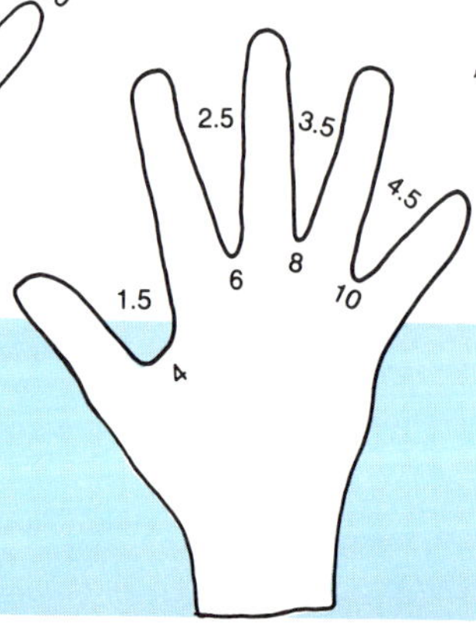

Figure 3.30

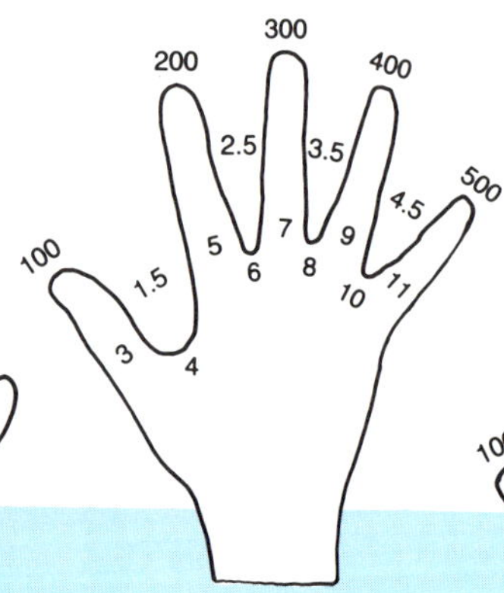

Figure 3.31

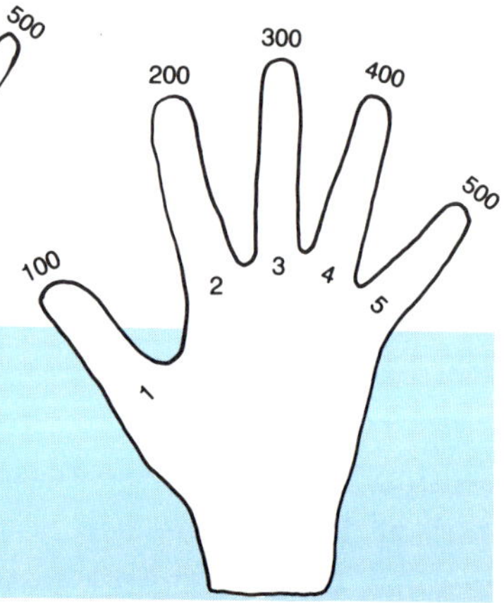

Figure 3.32

APPLYING RULES OF THUMB

To be proficient in the use of rule-of-thumb methods for quickly calculating fire stream problems an individual should study those that may be recommended, try them out and select one that applies to the individual's situation. After a rule-of-thumb method has been selected, it must be practiced until it can be used without difficulty.

The following layouts are presented as hypothetical problems to aid individuals in making their selection. The problems are solved by each of the rule-of-thumb methods. There is a difference in answers in some cases, but this difference is of little consequence for fireground operations: A few pounds per square inch difference, plus or minus, is insignificant. (For instance, pressure gauges can be inaccurate and difficult to read, but even such approximations aid the pump operator.) The following problem layouts may be made at the drill field and checked for accuracy with rule-of-thumb methods.

Example 33

What is the friction loss in a single 2½-inch hoseline 300 feet long with a 1-inch tip operating at 50 psi?

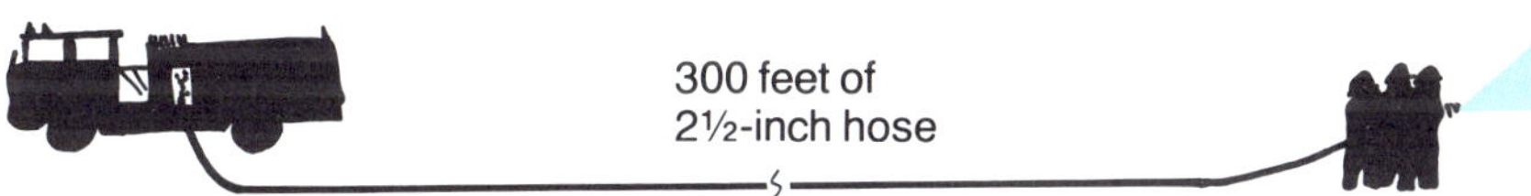

Figure 3.33

By GPM Flowing

1-inch tip flows 200 gpm at 50 psi. Subtract 10 from first two numbers of gpm: 20 − 10 = 10 psi. FL in 300 feet of 2½-inch hose = 3(10) = 30 psi.

By Counting

1-inch tip flows 200 gpm at 50 psi. Index finger represents 200 gpm at tip and 5 at base: 2(5) = 10 psi. FL in 300 feet of 2½-inch hose = 3(10) = 30 psi.

Example 34

What is the friction loss in a single 2½-inch line 450 feet long with a 1⅛-inch tip flowing at 50 psi?

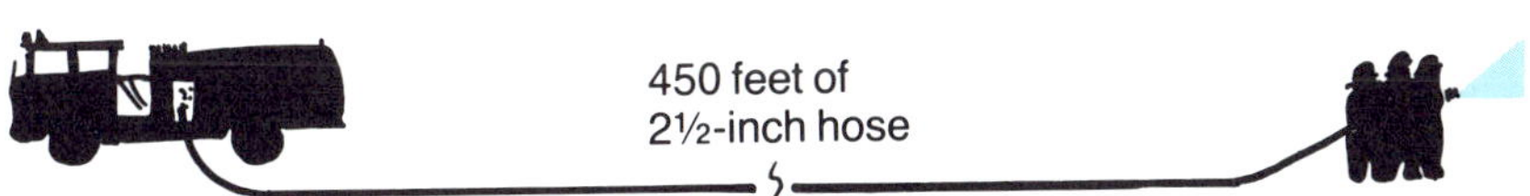

Figure 3.34

By GPM Flowing

Flow = 265 gpm. Subtract 10 from 26 = 16 psi. FL in 450 feet of 2½-inch hose = 4.5(16) = 72 psi.

By Counting

1⅛-inch tip flows 265 gpm at 50 gpm. The closest application is the space between index and second finger, which represents 250 gpm, and the matching number is six; thus, 2.5(6) = 15 psi. FL in 450 feet of 2½-inch hose = 4.5(15) = 67.5 psi.

NOTE: Any variation in the results obtained by the above methods should be considered when selecting a rule of thumb for fireground operations.

Example 35

What will the friction loss be in 400 feet of 2½-inch hose when using a 250 gpm fog nozzle at 100 psi?

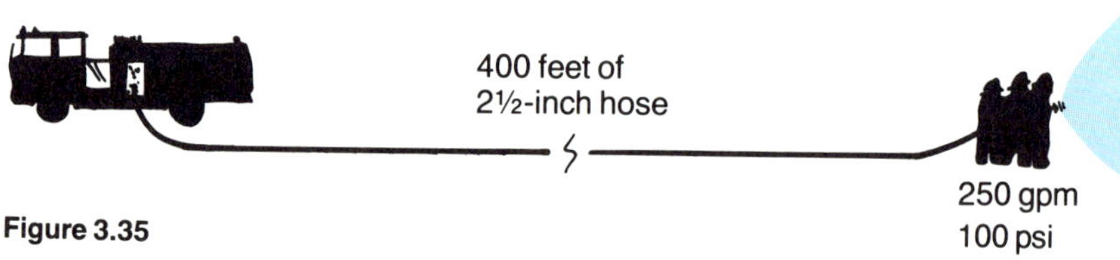

Figure 3.35

By GPM Flowing

Flow = 250 gpm. Refer to Table 3.15 where FL = 15 FL in 400 feet of 2½-inch hose = 4(15) = 60 psi.

By Counting

Between the index and middle finger: 2.5(6) = 15 psi per 100 feet. FL in 400 feet of 2½-inch hose = 4(6) = 60 psi.

Example 36

What will the friction loss be in two 2½-inch hoselines 300 feet long with 250 gpm flowing through each line? (The FL in both lines will be the same.)

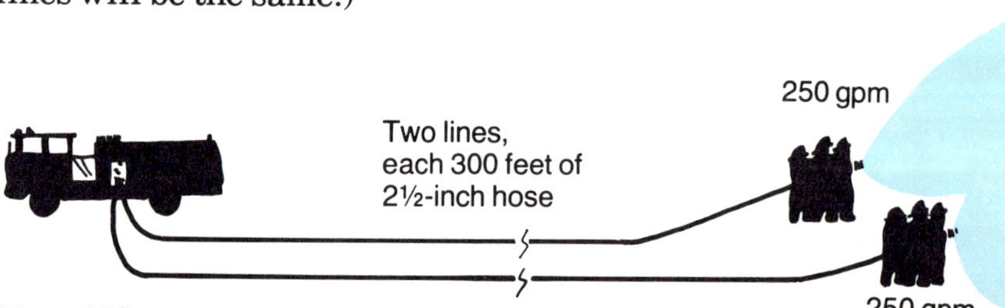

Figure 3.36

By GPM Flowing

Flow = 250 gpm. Subtract 10 from 25 = 15 psi.
FL in 300 feet of 2½-inch hose = 3(15) = 45 psi.

By Counting

Between the index and big fingers:
2.5(6) = 15 psi per 100 feet.
FL in 300 feet of 2½-inch hose = 3(15) = 45 psi.

Example 37

Find the desired pump discharge pressure for each of two 2½-inch hoselines, one 400 feet long and flowing 250 gpm, the other 600 feet long and feeding a 1-inch tip at 50 psi.

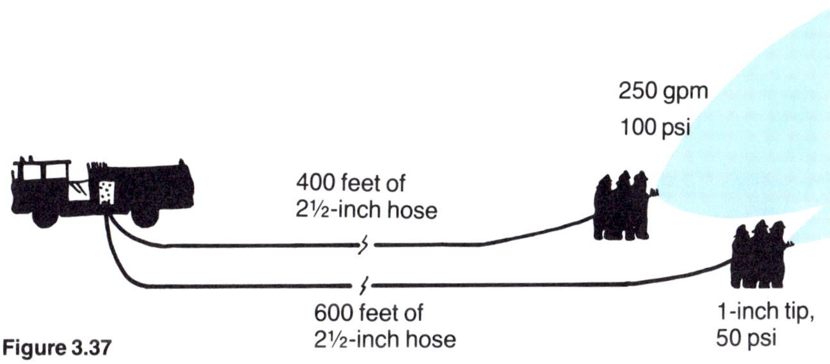

Figure 3.37

By GPM Flowing

400-foot line	600-foot line
Table 3.15 shows 15 psi FL	1-inch tip at 50 psi flows 200 gpm
FL in 400 feet = 4(15) = 60 psi	Subtract 10 from 20 = 10 psi
$PDP = NP + FL$	FL in 600 feet = 6(10) = 60 psi
$PDP = 100 + 60$	$PDP = NP + FL$
$PDP = 160$ psi	$PDP = 50 + 60$
	$PDP = 110$ psi

Example 38

What will be the FL in 500 feet of 3-inch hose with a 1¼-inch tip at 50 psi?

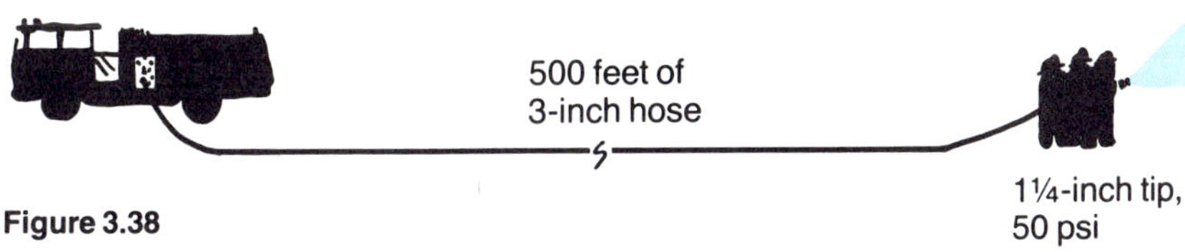

Figure 3.38

By Condensed Q Formula

1¼-inch tip at 50 psi delivers 300 gpm.

$$Q = \frac{gpm}{100}$$

$$Q = \frac{300}{100}$$

$$Q = 3$$

$FL = Q^2$

$FL = 3^2$

$FL = 9$ psi per 100 feet of 3-inch hose

FL in 500 feet of 3-inch hose = 5(9) = 45 psi

By GPM Flowing

Flow = 326 gpm

Subtract 10 from 32 = 22

FL for 3-inch hose = 22 divided by 2 = 11 psi per 100 feet

FL in 500 feet of 3-inch hose = 5(11) = 55 psi

By Counting

Q = 300 gpm approximately

Use middle finger: 3(3) = 9 psi per 100 feet of 3-inch hose

FL in 500 feet of 3-inch hose = 5(10) = 50 psi

Example 39

What will the FL be in each of two 3-inch hoselines 500 feet long with 250 gpm flowing through each line? (The FL in both lines will be the same.)

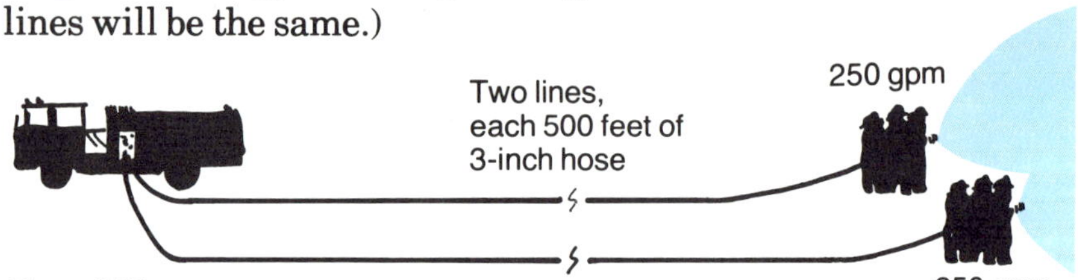

Figure 3.39

$$Q = \frac{gpm}{100}$$

$$Q = \frac{250}{100}$$

$$Q = 2.5$$

$FL = Q^2$

$FL = 2.5^2$

$FL = 6.25$ psi per 100 feet of 3-inch hose

FL in 500 feet of 3-inch hose = 5(6.25) = 31.25 psi

By GPM Flowing

Flow = 250 gpm

Subtract 10 from 25 = 15

FL for 3-inch hose = 5 divided by 2 = 7.5 psi per 100 feet

FL in 500 feet of 3-inch hose = 5(7.5) = 37.5 psi

By Counting

Q = 250 gpm

Use between index and middle finger

2.5(6) = 15 psi per 100 feet of 2½-inch hose

FL for 3-inch hose = 15 divided by 2 = 7.5 psi per 100 feet

FL in 500 feet of 3-inch hose = 5(7.5) = 37.5 psi

Example 40

What will be the total pressure loss (friction and elevation) from the following layout?

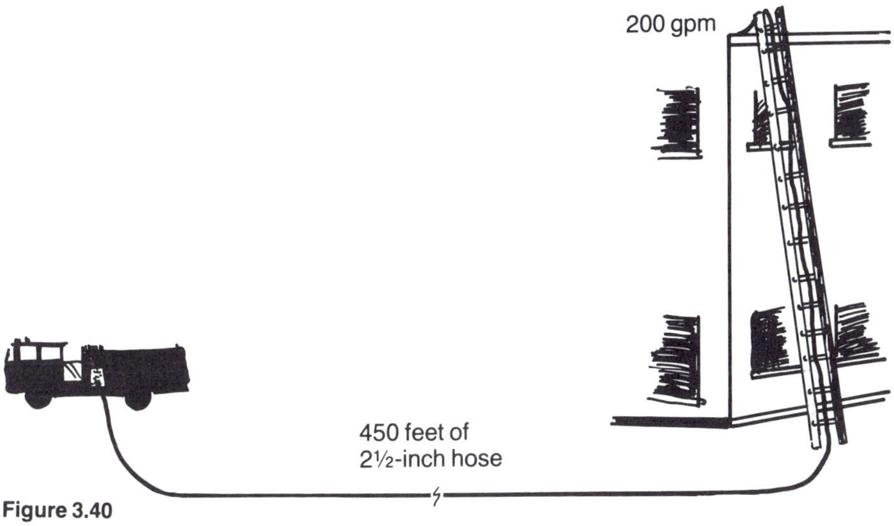

Figure 3.40

By GPM FLowing

Flow = 200 gpm

Subtract 10 from 20 = 10 psi FL per 100 feet

FL in 450 feet of 2½-inch hose = 4.5(10) = 45 psi

Loss due to elevation = 2(5) = 10 psi loss

Total loss = 45 + 10 = 55 psi

By Counting

Use index finger (200 gpm)

2(5) = 10 psi FL per 100 feet

FL in 450 feet of 2½-inch hose = 4.5(10) = 45 psi

Loss due to elevation = 2 stories times 5 psi per story = 10 psi loss

Total loss = 45 + 10 = 55 psi

Example 41

In the following illustration, what will be the pressure "gain" and the friction loss due to a depression?

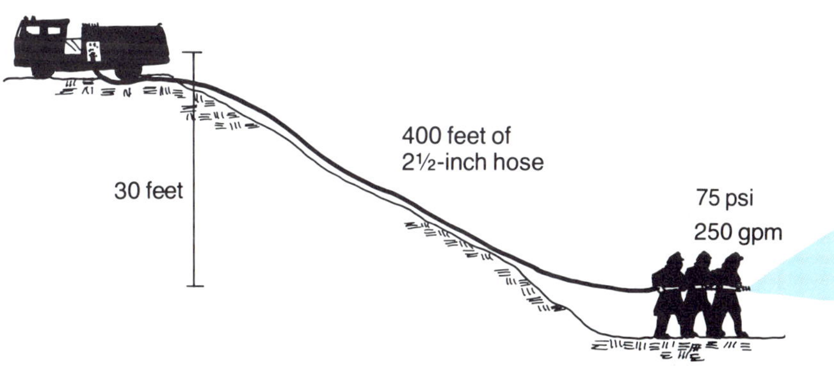

Figure 3.41

By GPM Flowing
Flow = 250 gpm
FL = 15 psi per 100 feet
Total FL = 4(15) = 60 psi

Pressure "Gain"
0.5 psi per foot
Pressure gain = 0.5(30) = 15 psi

Total loss = 60 − 15 = 45 psi

USING IN-LINE GAUGES TO FIND FRICTION LOSS

An in-line gauge can be used to find the approximate friction loss in a hoseline. Figure 3.42 shows an easy, workable layout for such a test.

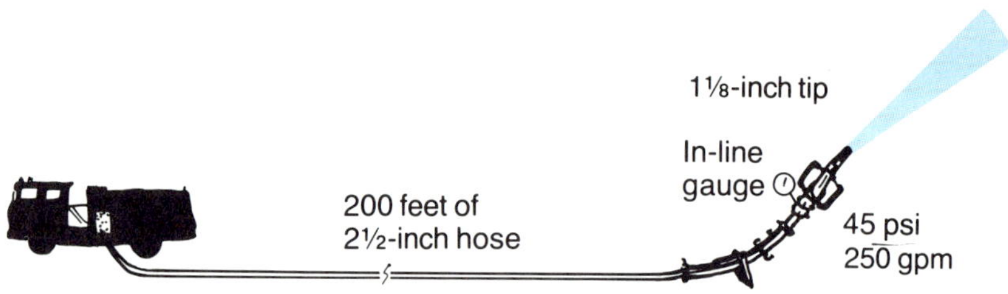

Figure 3.42

A solid stream nozzle, rather than a fog nozzle, should be used because the pressure will be lower and safer, and the nozzle and line easier to control. The reading of the in-line gauge is subtracted from the pump discharge reading to give the friction-loss figure for the entire length of hose. The in-line gauge reading is also used to find the flow in gpm.

Using Figure 3.42 as an example, the friction loss in the hose being tested would be determined as follows. First, the *desired* flow is determined; in this case, 250 gpm. By mathematically rearranging the formula for gpm flow, the nozzle pressure needed to expel 250 gpm can be found:

$$P = \frac{gpm}{29.7d^2}$$

$$P = \frac{250}{29.7(1.125^2)}$$

$$P = \frac{250}{29.7(1.27)}$$

$$P = \frac{250}{37.7}$$

$$P = 6.63^2$$

$$P = 43.95$$

FLOW METERS

Flow meters mounted in the pump operator's panel are great aids for producing fire streams. They are very simple mechanisms plumbed directly into the discharge system. A flow meter automatically registers the flow from the nozzle. This frees the pump operator from having to make friction-loss calculations. All the pump operator has to know is the rated discharge of the nozzle, then all that has to be done is to adjust the engine throttle until the desired flow is shown on the flow meter.

A test conducted at the Hartland Fire Training Center, San Diego County, California, underscores the usefulness of flow meters. In the test, four portable monitors of the same design were supplied with water, one after the other, by the same pumper and the same hose. The engine speed was increased until the flow meter read 1,000 gpm. The necessary pump discharge pressures for the four monitors varied: 117, 132, 109 and 105 psi — a high-low difference of 27 psi. This experiment points out that even "correct" calculations can be inadequate for determining the correct pressure for the desired flow. The friction loss in appliances of the same make and model can vary.

CALCULATIONS FOR RELAY PUMPING

Relay pumping is used when the distance between the water source and the fire is too great for one pumper to overcome the friction loss. Relays are also used when the back pressure of elevated lines is too much for one pumper.

Spacing Pumpers

The number of pumpers needed and the distances between them is determined by several factors:

- Amount of water needed at the fire
- Distance between water source and fire
- Hose size
- Pumper capacity
- Terrain

The first determination to be made in spacing the pumpers for a relay is the number of pumpers needed for the relay. This depends on the maximum allowable working pressure, the distance that hose must be laid, the size(s) of the hose available, and the capacities of the pumpers.

If the maximum allowable working pressure is 200 psi (based on a hose test of 250 psi less 50 psi for a safety margin), the available engine pressure (*EP*) is 180 psi (20 psi subtracted for necessary residual pressure at the intake side of the next pumper). This 180 psi is the pressure available to overcome friction loss in the hose between pumpers. Table 3.17 lists the distances that 180 psi will deliver at various flows through 2½-, 3-, 4-, 5- and 6-inch hose. The number of pumpers needed for a relay can be found by using this table.

TABLE 3.17
Maximum Distance of Waterflow At 180 PSI
(in feet)

Flow (U.S. gpm)	Hose Diameter (inches) 2½	3	4	5	6
100	6,000	18,000			
200	1,800	4,500	18,000		
250	1,200	3,000	12,000		
300	850	2,000	9,000	27,000	
400	500	1,100	4,500	16,000	
500	300	720	3,000	10,500	30,000
750	150	350	1,500	4,800	12,000
1,000		200	900	2,700	6,500

NOTE: Figures rounded down to allow for variable terrain.

Figures determined by the formula $Distance = \frac{\text{psi}(100)}{FL/100\ feet}$

Example 42

If a single line of 3-inch hose is used, how many pumpers will be needed to supply a 1,000 gpm master stream device that is 1,000 feet from the water source?

From the table, the maximum distance that water will flow at 1,000 gpm through 3-inch hose is 200 feet. Divided this figure into the distance to the fire; then add one for the attack pumper:

$$\frac{1{,}000}{200} = 5 + 1 = 6 \text{ pumpers needed}$$

Example 43

If two lines of 3-inch hose are used, how many pumpers will be needed to supply a 1,000 gpm master stream device that is 2,000 feet from the water source?

First divide the flow in half (500). Find the distance in the Table for 3-inch hose at 500 gpm (720 feet). Then divide this distance into the total distance.

$$\frac{2{,}000}{720} = 2.7 + 1 \text{ (for attack pumper)} = 3.7\text{, or 4 pumpers needed}$$

Exercise 20

If one line of 4-inch hose is used, how many pumpers are needed for a relay 2,000 feet long supplying a 500 gpm master stream device?

The number of pumpers necessary for the preceding examples is adequate if the pumpers have the ability to deliver 1,000 gpm at 180 psi. The amount of water that a pumper can deliver decreases with each increase in pressure. Put another way, an increase in volume decreases the pressure that the pump can develop. Table 3.18 shows this relationship between volume and pressure.

TABLE 3.18
Maximum Pressure/Volume of Class A Pumpers

Percent of Capacity	Pressure (psi)
100	150
70	200
50	250

From Table 3.18, one finds that a pumper rated at 1,000 gpm will deliver 1,000 gpm at 150 psi, 700 gpm at 200 psi and 500 gpm at 250 psi. The pumpers for Examples 42 and 43, then, would have to be rated at 1,500 gpm to deliver 1,000 gpm at 180 psi. If 1,500 gpm pumpers were not available, more pumpers would be needed. The pumpers for Exercise 20 would have to be rated at 750 gpm or more.

Usually, the distance between pumpers is not a critical matter. Most pumpers in a relay will have laid all except 100 feet of their hose, and most pumpers do not carry more hose of any one diameter than they can supply. (The 100 feet held in reserve is in case a hose bursts or needs to be replaced.) Experience and practice will help companies in making the determinations of what sizes of hose to use and where to position the pumpers in a relay.

A Rule of Thumb

The following rule of thumb has been devised for determining the position of the second pumper in a two-pumper relay using 2½-inch hose.

Step 1: Know the distance between the fire and the water source.

Step 2: Find the friction loss in 2½-inch hose per 100 feet for the required flow.

Step 3: Divide the required nozzle pressure by the friction loss per 100 feet. The answer is imaginary hundreds of feet added to the real distance.

Step 4: Divide the total of the real and imaginary distance by 2. This answer gives the distance from the water source that the second pumper should be positioned.

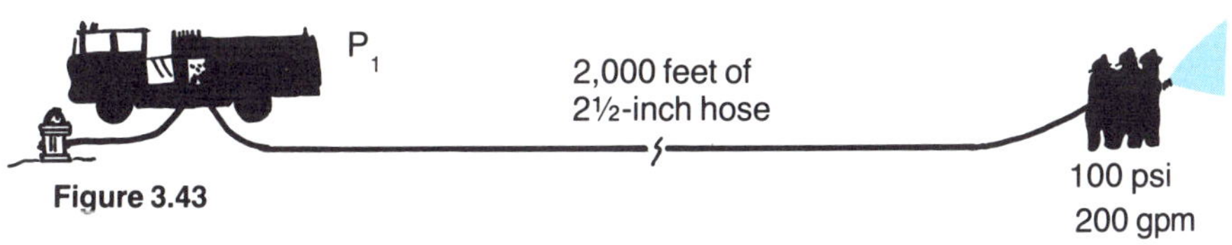

Figure 3.43

Example 44

Where should the second pumper be positioned?

Figure 3.44

Step 1: The lay is 1,000 feet long.

Step 2: FL in 100 feet of 2½-inch hose, at 200 gpm, is 10 psi.

Step 3: $\frac{NP}{FL} = \frac{100}{10} = 10$

Add 10 hundreds (1,000) feet to the actual distance: 3,000 feet.

Step 4: Divide 3,000 by 2.

The second pumper should be located 1,500 feet from the first pumper.

Exercise 21

Where should the second pumper be positioned?

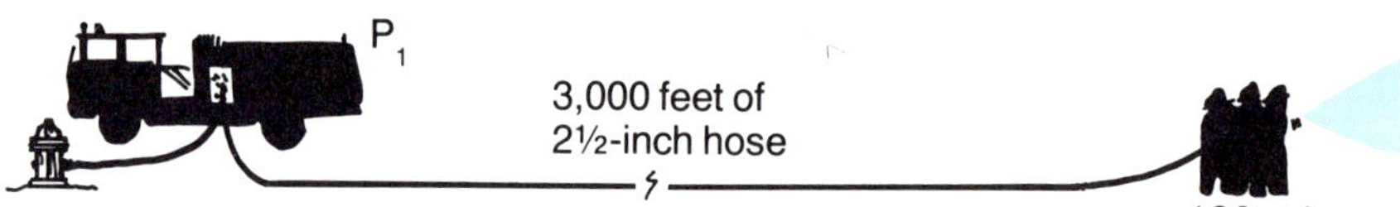

Figure 3.45

The formula for the required pump discharge pressure for each pumper in a relay is basically the same as for a single-pumper layout:

$$PDP = FL + RP + EL$$

where

PDP = Pump discharge pressure (psi)

FL = Friction loss

RP = Residual pressure for the next pumper (20 psi)

EL = Pressure "loss" or "gain" from elevation

The pump discharge pressure for the attack pumper is calculated with the standard formula:

$$PDP = FL + NP + EL$$

Example 45

What is the required pump discharge for each of the pumpers in this relay layout?

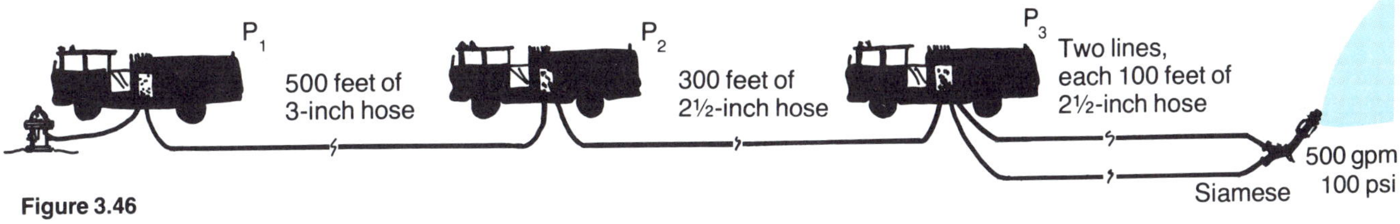

Figure 3.46

1. Determine the PDP for Pumper 3:

$$PDP = FL + NP + EL$$

FL in hose = 15.2 psi
FL in device = 10 psi
Total FL = 25 psi (rounded off)

$$PDP = 25 + 100 + 0$$
$$PDP = 125$$

2. Determine the PDP for Pumper 2:

$$PDP = FL + RP + EL$$

FL = 55 psi per 100 feet
FL = 55(3) = 165 psi

$$PDP = 165 + 20 + 0$$
$$PDP = 185 \text{ psi}$$

3. Determine the PDP for Pumper 1:

$$PDP = FL + RP + EL$$

FL = 25 psi per 100 feet
FL = 25(5) = 125 psi

$$PDP = 125 + 20 + 0$$
$$PDP = 145 \text{ psi}$$

Exercise 22

What is the required pump discharge pressure for each of these pumpers?

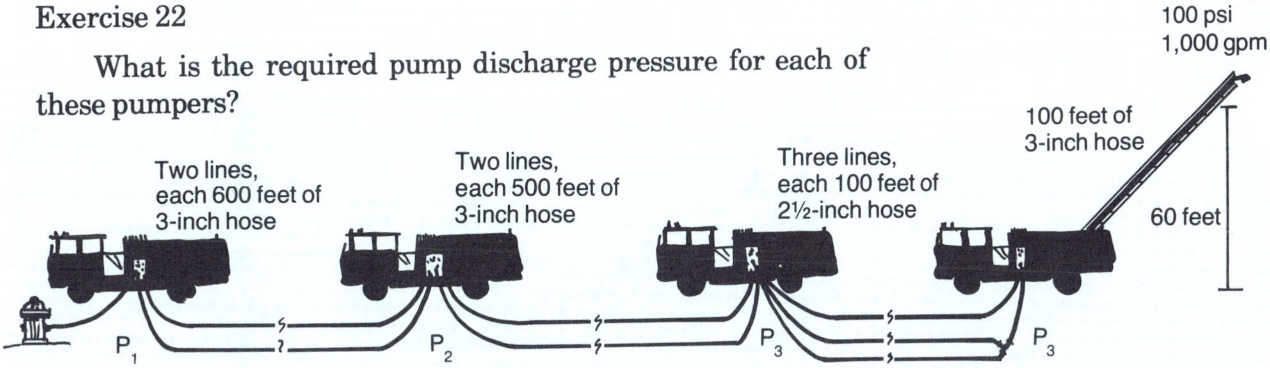

Figure 3.47

Exercise 23

What is the required pump discharge pressure for each of these pumpers?

Figure 3.48

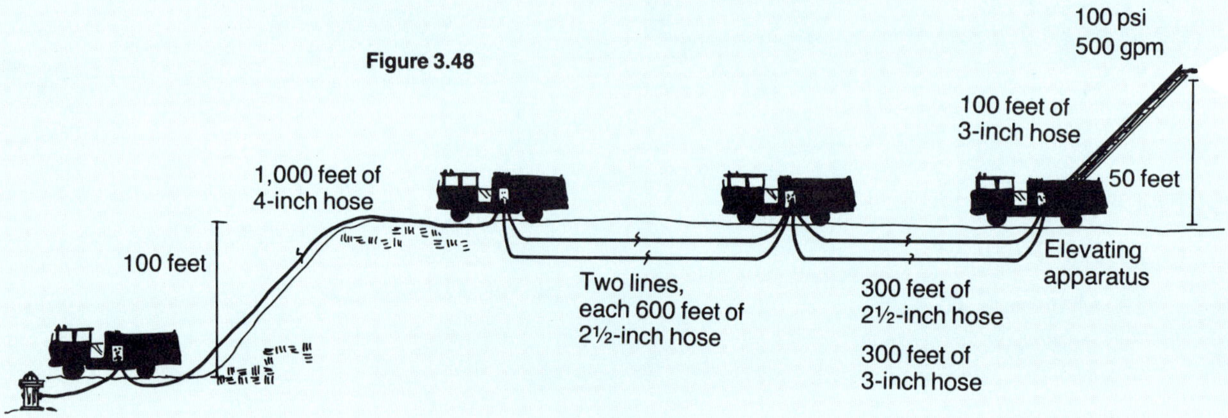

Forming the Relay

- First-arriving engine is positioned at the fire. The person in charge of this company makes the size-up and determines the quantity of water needed. An initial attack may be made with the water carried on the pumper.
- When the relay is being set up, the largest capacity pumper available should be positioned at the water source.
- The distance between each pumper is determined. Pumpers should keep at least 100 feet of hose in reserve in case a hoseline bursts.
- The pump operator for each pumper except the supply pumper opens an unused discharge gate if the pump does not have a relay valve. This allows the air from the hoselines to escape as the water advances up the hoseline.
- The pumper at the water source pumps water at 150 psi.
- The pump operator at the next pumper closes the unused discharge gate when water flows from it, then advances the throttle until 150 psi is developed. Each successive pump operator follows the same procedure.
- The operator of the attack pumper informs the other pump operators how much water is needed.
- The supply pumper throttle is adjusted until the operating pressure is reached for the needed amount of water. Each successive pump is similarly adjusted. The discharge pressure should not exceed 200 psi if the hose has been tested at 250 psi. The residual pressure reading at the intake gauge should not fall below 20 psi.

- Set pressure regulating devices.
- The attack-pump operator adjusts the discharge pressure(s) to supply the attack line(s).

Once the water is moving, the supply should not be shut off at any point between the supply pumper and the fire. Nozzles should not be closed unless absolutely necessary. If a temporary shut-down is called for, the attack-pump operator should open one or more discharge gates to dump excess water. If a hoseline bursts, a discharge gate of the pumper before the rupture should be opened to dump water until the length is replaced.

Shutting Down

- Do not close nozzles. Attack pumper is throttled down but water is kept discharging until the pump can be disengaged. (Pumps in relay are in effect a multiple-stage pump. Enormous pressures can be built up.)
- Successively, the pumpers toward the water source follow the same procedure. An unused discharge gate is opened on each pumper.
- The water supply pumper is shut down.

DUAL PUMPING OPERATIONS: HYDRANT

When dual pumping (also known as tandem pumping), one hydrant may be used to supply two pumpers if its volume is equal to or greater than the combined rated capacity of the pumpers. This type of operation has several advantages, including better use of available water, and shorter hose lays (particularly if the hydrant is close to the fire). Additional hoselines can be placed in operation quicker, and apparatus may be grouped close together, allowing easier coordination. The method for a tandem pumping hook-up is as follows:

Step 1: Pumper 1 connects to the hydrant steamer connection, using large intake hose. This pumper then pumps water through its lines to the fire.

Step 2: Pumper 2 is positioned intake-to-intake with Pumper 1. The hydrant is closed down until the compound gauge of Pumper 1 reads near 0 (about 5 psi). The throttle of Pumper 1 will need adjusting. This makes the volumes of discharge and intake equal, so the cap of the unused intake can be removed.

Step 3: Pumper 2 is connected by large intake hose to the unused steamer intake of Pumper 1.

Step 4: The hydrant is opened completely.

Step 5: Pumper 2 pumps water through its lines to the fire. Its supply is the water available that is not being used by and passing through Pumper 1.

SUPPLYING PRIVATE FIRE PROTECTION EQUIPMENT

Many properties have private fire protection equipment provided to protect lives and property. Fire protection equipment of this nature, in addition to portable fire extinguishers and fixed extinguishing installations, includes sprinkler and standpipe systems. Each year, firefighters respond to thousands of alarms in properties having such protection. For private protection equipment to function as it should, fire department personnel must make pre-fire plans to use the equipment properly (see IFSTA 302, **Fire Department — Facilities, Planning and Procedures**.)

Supplying Automatic Sprinkler Systems

An automatic sprinkler system can be of great assistance to a fire department if adequate fire companies support the system. Records show that fire departments generally make good use of these systems, but there have been large-loss fires because the department did not effectively use this valuable fire fighting tool. Such occurrences reflect the need for fire department personnel to plan support operations for sprinklered properties, including a very complete inspection and survey. When a fire occurs, the plan of operation should be followed as far as possible.

Some of the essentials for adequate support of private fire protection equipment are as follow:

- Make sure that the pumper is of adequate volume and pressure.
- Use two hoselines from the pumper to the fire department connection.
- Build up pressure slowly to a minimum of 100 psi (700 kPa) — pressure must be greater than static pressure of the system. If one pumper is unable to do this, use two pumpers.
- Do not rob the water supply feeding the system.
- Check all control valves to the system.
- Avoid premature shutoff of sprinkler system.
- Maintain the pumper hookup during overhaul.
- See that the system is restored after the fire is known to be out.

NOTE: The pressures needed for each system should be part of the pre-fire plan because of variances in system and age.

A more complete study of types of sprinkler systems, water supplies, layouts and control valves can be found in IFSTA 210, **Private Fire Protection and Detection.**

Feeding Standpipe Systems

Class I systems (2½-inch [65 mm] hose connections) are provided for use by fire departments and those trained in handling heavy fire streams.

Class II systems (1½-inch [38 mm] hose connections) are provided for use primarily by the building occupants until the fire department arrives.

Class III systems are provided for use by fire department personnel and those trained in heavy hose streams or by the occupants.

For fire department use, standpipe systems should deliver at least 500 gpm (1,875 L/min) for a period of at least 30 minutes where only one standpipe is required. Where more than one standpipe is required, the minimum supply is 500 gpm for the first standpipe and 250 gpm (940 L/min) for each additional standpipe, with the total supply not exceeding 2,500 gpm (9,375 L/min) for at least 30 minutes. In addition, the supply is required to be strong enough to maintain a residual pressure of 65 psi (450 kPa) at the top outlet for each standpipe with 500 gpm flowing.

The water supply for Class II service is 100 gpm (375 L/min) for a period of at least 30 minutes. It is also required to be strong enough to maintain a residual pressure of 65 psi at the top outlet, including roof outlets with 100 gpm flowing. Connection for a public fire department to pump water into a standpipe system is a requirement for Class I and III systems.

Some properties have standpipe systems that serve fire hose outlets in various parts of one or more buildings. Standpipe systems of various types should be used by the fire department to quickly place streams in service in areas that cannot be conveniently reached with hoselines from pumpers.

Fire department operations in buildings where standpipe systems supply fire department hose streams require procedures as carefully planned as those for buildings protected by automatic sprinklers.

When standpipe systems are provided with fire department connections a pumper should supply water through parallel hoselines. To insure effective fire streams on the various floors the following items or factors must be considered and used.

- A pump operator should always allow 25 psi (170 kPa) loss for the standpipe system. This 25 psi loss allowance covers the fire department connection, clapper valve obstruction, pipe bends and hose connections.
- The usual allowance of 5 psi (35 kPa) for each floor because of elevation or 0.5 psi (3.0 kPa) per foot of elevation must be considered.
- The friction loss per 100 feet (30 m) of hose must be based upon the gpm flow. The friction loss in hose will vary with the length of working line at the fire area and the length of the two parallel supply hoselines.

The use of these factors will provide a fairly accurate system for calculating both solid and fog fire streams from standpipe systems.

Example 46

100 feet of 2½-inch hose stretched from riser on 10th floor to fire area. 200 gpm fog nozzle at 100 psi. Two parallel 2½-inch lines 200 feet long from pumper to standpipe siamese connection.

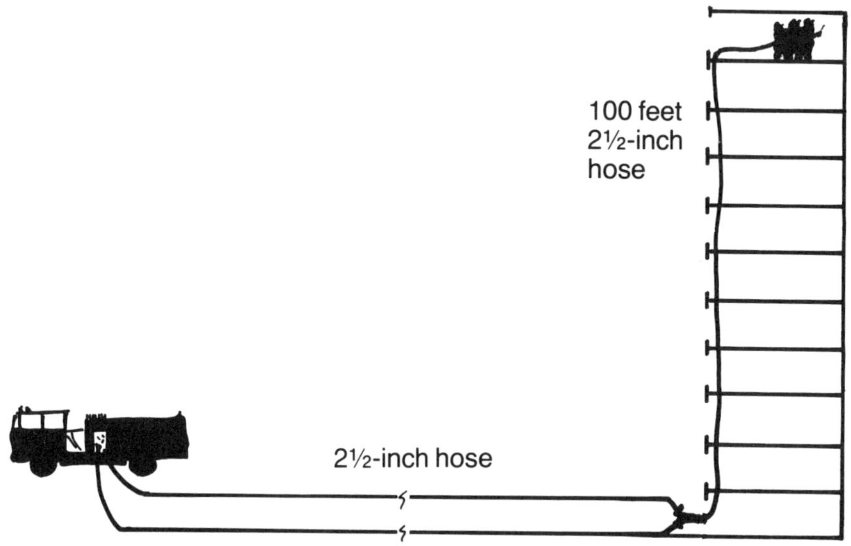

Figure 3.49

NP	=	100 psi
100 feet of 2½-inch hose	=	10 psi
10 floors elevation above 1st floor	=	50 psi
Standpipe system	=	25 psi
200 feet parallel lines	=	6 psi
PDP	=	191 psi

ANSWERS TO EXERCISES

Exercise 1

25
$3(4) = 12$
1,225
12.25

Exercise 2

25
$72(73) = 5{,}256$
525,625
52.5625

Exercise 3

25
$5(6) = 30$
3,025

Exercise 4

$$A = \pi r^2$$
$$A = 3.1416(0.75^2)$$
$$A = 3.1416(0.5625)$$
$$A = 1.76715$$

$$A = \frac{\pi d^2}{4}$$
$$A = \frac{3.1416(1.5^2)}{4}$$
$$A = \frac{3.1416(2.25)}{4}$$
$$A = \frac{7.0686}{4}$$
$$A = 1.76715$$

Exercise 5

$$A = \pi r^2$$
$$A = 3.1416(15^2)$$
$$A = 3.1416(225)$$
$$A = 706.86$$

$$A = \frac{\pi d^2}{4}$$
$$A = \frac{3.1416(30^2)}{4}$$
$$A = \frac{3.1416(900)}{4}$$
$$A = \frac{2827.44}{4}$$
$$A = 706.86$$

Exercise 6

$$ft^3 = \pi r^2 L$$
$$ft^3 = 3.1416(5^2)(30)$$
$$ft^3 = 3.1416(25)(30)$$
$$ft^3 = 78.54(30)$$
$$ft^3 = 2{,}356.2$$
[7.48 gal = 1 ft^3]

gallons = 2,356.2(7.48)
gallons = 17,624.376

(short method)
$$gal = d^2 6L$$
$$gal = 10^2(6)(30)$$
$$gal = 100(6)(30)$$
$$gal = 600(30)$$
$$gal = 18{,}000$$

Exercise 7

$ft^3 = \pi r^2$
$ft^3 = 3.1416(1.5^2)(10)$
$ft^3 = 3.1416(2.25)(10)$
$ft^3 = 7.0686(10)$
$ft^3 = 70.686$

$gal = 70.686(7.48)$
$gal = 528.73$

$gal = d^2 6L$
$gal = 3^2(6)(10)$
$gal = 9(6)(10)$
$gal = 54(10)$
$gal = 540$

Exercise 8

$gpm = 29.7d^2 \sqrt{P}$
$gpm = 29.7(2^2)(\sqrt{80})$
$gpm = 29.7(4)(8.9443)$
$gpm = 29.7(35.7772)$
$gpm = 1{,}063$ (rounded off)

Exercise 9

$gpm = 29.7d^2 \sqrt{P}$
$gpm = 29.7(1.125^2)(\sqrt{60})$
$gpm = 29.7(1.266)(7.7460)$
$gpm = 29.7(9.8)$
$gpm = 291$ (rounded off)

Exercise 10

$Q = \frac{gpm}{100}$

$Q = \frac{250}{100}$

$Q = 2.5$

$FL = 2Q^2 + Q$
$FL = 2(2.5^2) + 2.5$
$FL = 2(6.25) + 2.5$
$FL = 12.5 + 2.5$
FL = 15 psi per 100 feet of 2½-inch hose

Exercise 11

$Q = \frac{400}{100}$

$Q = 4$

$FL = 2Q^2 + Q$
$FL = 2(4^2) + 4$
$FL = 2(16) + 4$
$FL = 32 + 4$
FL = 36 psi per 100 feet of 2½-inch hose

Exercise 12

$$Q = \frac{300}{100}$$

$$Q = 3$$

$FL = 2Q^2 + Q$
$FL = 2(3^2) + 3$
$FL = 2(9) + 3$
$FL = 18 + 3$
FL = 21 psi per 100 feet of 2½-inch hose
FL in 500 feet is 5(21) = 105 psi

Exercise 13

$gpm = 29.7d^2\sqrt{P}$
$gpm = 29.7(1^2)(\sqrt{65})$
$gpm = 29.7(1)(8)$ [nearest square root]
$gpm = 29.7(8)$
$gpm = 238$

$FL = 2Q^2 + Q$
$FL = 2(2.4^2) + 2.4$
$FL = 2(5.76) + 2.4$
$FL = 11.52 + 2.4$
FL = 13.92 or 14 psi per 100 feet of 2½-inch hose
FL in 400 feet is 4(14) = 56 psi

$$Q = \frac{gpm}{100}$$

$$Q = \frac{238}{100}$$

$Q = 2.4$ (rounded off)

Exercise 14

$$Q = \frac{gpm}{100}$$

$$Q = \frac{250}{100}$$

$$Q = 2.5$$

$FL = 2Q^2 + Q$
$FL = 2(2.5^2) + 2.5$
$FL = 2(6.25) + 2.5$
$FL = 12.5 + 2.5$
FL = 15 psi per 100 feet of 2½-inch hose
FL in 250 feet is 2.5(15) = 37.5 psi

$PDP = NP + FL$
$PDP = 100 + 37.5$
$PDP = 137.5$ psi

Exercise 15

(1-inch tip, 50 psi = 200 gpm)

$$Q = \frac{gpm}{100}$$

$$Q = \frac{200}{100}$$

$$Q = 2$$

$FL = 2Q^2 + Q$
$FL = 2(2^2) + 2$
$FL = 2(4) + 2$
$FL = 8 + 2$
FL = 10 psi per 100 feet of 2½-inch hose

FL in 200 feet = 20 psi
FL in 300 feet = 30 psi
$PDP = NP + FL$
$PDP = 50 + 30$ (highest FL)
$PDP = 80$ psi

Exercise 16

First find FL of the 1½-inch hose:

$$Q = \frac{gpm}{100} \qquad FL = 2Q^2 + 0.5(Q)$$

$$Q = \frac{95}{100} \qquad FL = 2(0.95^2) + 0.5(0.95)$$

$$FL = 2(0.9) + 0.47$$

$$FL = 1.8 + 0.47$$

$$Q = 0.95 \qquad FL = 2.27 \text{ psi per 100 feet of 2½-inch hose}$$

Conversion factors from Table 3.2 indicate multiplication by 13.5 or division by 0.074.

$$42.27\ (0.074) = 30.645 \text{ psi}$$

$$\frac{2.27}{13.5} = 30.67 \text{ psi}$$

Short line FL in 150 feet = 1.5(30.6) = 45.9 psi
Long line FL in 200 feet = 2(30.6) = 61.2 psi

The flow in the 2½-inch hose is 2(95) = 190 gpm. Find the FL for the 2½-inch hose.

$$Q = \frac{gpm}{100} \qquad FL = 2Q^2 + Q$$

$$FL = 2(1.9^2) + 1.9$$

$$Q = \frac{190}{100} \qquad FL = 2(3.61) + 1.9$$

$$FL = 7.2 + 1.9$$

$$Q = 1.9 \qquad FL = 9.1 \text{ (or 9) psi per 100 feet of 2½-inch hose}$$

FL in 200 feet = 2(9) = 18 psi

Using these figures, the total friction loss is:

61.2 psi	FL in longest 1½-inch line
5.0 psi	FL in wye
+ 18.0 psi	FL in 2½-inch line
84.2 psi	Total FL

Exercise 17

$$Q = \frac{gpm}{100} \qquad FL = 2Q^2 + Q$$

$$FL = 2(2^2) + 2$$

$$Q = \frac{200}{100} \qquad FL = 2(4) + 2$$

$$FL = 8 + 2$$

$$Q = 2 \qquad FL = 10 \text{ psi per 100 feet of 2½-inch hose}$$

FL in 400 feet of 2½-inch hose = 4(10) = 40 psi

The flow of water through the 3-inch hose supply line will be the total flow of both 2½-inch hoselines and fog nozzles, or 400 gpm.

$$gpm = 2(200) = 400$$

$$Q = \frac{gpm}{100} \qquad Q = \frac{400}{100} \qquad Q = 4$$

$$FL = 2Q^2 + Q$$
$$FL = 2(4^2) + 4$$
$$FL = 2(16) + 4$$
$$FL = 32 + 4$$
$$\text{FL} = 36 \text{ psi per 100 feet of 2½-inch hose}$$

Conversions factors for 3-inch hose with 2½-inch couplings from Table 3.2.

$$36(0.4) = 14.4 \text{ psi}$$

or

$$\frac{36}{2.5} = 14.4 \text{ psi}$$

FL in 500 feet of 3-inch hose = 5(14) = 70 psi

Total Friction Loss

40 psi	FL in 2½-inch hose
5 psi	FL in wye
70 psi	FL in 3-inch hose
115 psi	Total FL

$$PDP = NP + FL$$
$$PDP = 100 + 115$$
$$PDP = 215 \text{ psi}$$

Exercise 18

$$Q = \frac{gpm}{100} \qquad Q = \frac{95}{100} \qquad Q = 0.95$$

$$FL = 2Q + 0.5Q$$
$$FL = 2(0.95^2) + 0.5(0.95)$$
$$FL = 2(0.90) + 0.47$$
$$FL = 1.8 + 0.47$$
$$\text{FL} = 2.27 \text{ psi per 100 feet of 2½-inch hose}$$

Conversion factors for 1½-inch hose from Table 3.2.

$$2.27(13.5) = 30.645 \text{ psi}$$

$$\frac{2.27}{0.074} = 30.67 \text{ psi}$$

FL in 200 feet of 1½-inch hose = 2(30.6) = 61.2 psi

The flow of water through the 2½-inch hose supply line will be the total of all three 1½-inch hoselines and nozzle, or 285 gpm.

$$gpm = 3(95) = 285$$

$$Q = \frac{gpm}{100}$$

$$Q = \frac{285}{100}$$

$$Q = 2.85$$

$$FL = 2Q^2 + Q$$
$$FL = 2(2.85^2) + 2.85$$
$$FL = 2(8.12) + 2.85$$
$$FL = 16.24 + 2.85$$

FL = 19 psi per 100 feet of 2½-inch hose

FL in 300 feet of 2½-inch hose = 3(10) = 57 psi

Total Friction Loss

61 psi	FL in 1½-inch hose
5 psi	FL in 3-inch wye
57 psi	FL in 2½-inch hose
123 psi	Total FL

$$PDP = NP + FL$$
$$PDP = 100 + 123$$
$$PDP = 223 \text{ psi}$$

Exercise 19

$$gpm_2 = 0.1(\sqrt{NP_2})(gpm_1)$$
$$gpm_2 = 0.1(\sqrt{225})(30)$$
$$gpm_2 = 0.1(15)(30)$$
$$gpm_2 = 45.0$$

Exercise 20

$$\frac{2{,}000}{3{,}000} = 0.66 = 1 + 1 = 2 \text{ pumpers needed.}$$

Exercise 21

Step 1: The lay is 3,000 feet long.

Step 2: FL in 100 feet of 2½-inch hose, at 200 gpm, is 10 psi.

Step 3: $\frac{NP}{FL} = \frac{100}{10} = 10$

Step 4: Divide 3,000 by 2.

The second pumper should be located 2,000 feet from the first pumper.

Exercise 22

PDP for Pumper 4: $PDP = FL + NP + EL$

$FL = 80.9$

$PDP = 80.9 + 100 + 30$

$PDP = 211$ psi (rounded off)

PDP for Pumper 3: $PDP = FL + RP + EL$

$FL = 27.0$

$PDP = 27.0 + 20 + 0$

$PDP = 47$ psi

PDP for Pumper 2: $PDP = FL + RP + EL$

$FL = 21.2(5) = 106$

$PDP = 106 + 20 + 0$

$PDP = 126$ psi

PDP for Pumper 1: $PDP = FL + RP + EL$

$FL = 21.2(6) = 127.2$

$PDP = 127.2 + 20 + 0$

$PDP = 147$ psi (rounded off)

Exercise 23

PDP for Pumper 3: $PDP = FL_s + FL_{3\text{-inch}} + NP + EL + AL$

$FL_s = \frac{55}{6} = 9$ psi per 100 feet

$FL_s = 9(3) = 18$ psi per 300 feet

$FL_{3\text{-inch}} = 21.2(3) = 64.6$ psi per 300 feet

$PDP = 18 + 63.6 + 100 + 25 + 10$

$PDP = 216$ psi (rounded off)

PDP for Pumper 2: $PDP = FL + RP + EL$

$PDP = 15.2(6) + 20 + 0$

$PDP = 91.2 + 20$

$PDP = 110$ psi

PDP for Pumper 1: $PDP = FL + RP + EL$

$PDP = 5(10) + 20 + 50$

$PDP = 50 + 20 + 50$

$PDP = 120$ psi

METRIC SYSTEM OF MEASUREMENT

Since the development of the metric system in the eighteenth century the vast majority of the world's countries have converted to the metric system. In fact, by 1974 only six countries were uncommitted to the system.

There are two basic reasons for this conversion. The first is simplicity and efficiency. The metric system is based upon simple powers of ten, where mass, volume and length are all interrelated. Using the power of ten makes computations simple, with a reduction in errors and a saving of time. The second reason is fairly obvious: the simple economics of world trade demand a common system of measurement. The majority of countries use metrics; they supply products in metric measurements. These same countries, as consumers, often demand products in metrics. A country that wants to remain economically competitive in the world market must convert to the metric system.

The metric system used worldwide today, Le Systeme International d'Unites (simply called SI), was agreed upon by the metric countries in 1960.

Familiarity and ease with other systems of measurement are the result of a lifetime of experience. The metric system at first glance may seem much more complicated; however, once proficiency is gained in the use of the metric system the user will wonder why the conversion was not made years ago.

Metric equivalents are given in parentheses after English measures in this manual so firefighters who are accustomed to the English system may become familiar with them. The gallon is the U.S. gallon. To help simplify this introduction to metrics a soft approach has been taken using mostly even numbers rather than the exact figure (hard approach).

SI Rules

The metric system does have its rules, which are necessary especially because of its international character. They are few and easy to remember.

1. The short versions of the units are symbols in their own right, not abbreviations. No periods are placed after them except at the end of a sentence. The symbol for kilogram, for instance, is kg, not "kg.".

2. Symbols are always written in lower case with the exceptions of the first letter of a symbol for a unit named after an individual and the symbols for the three "biggest" prefixes: Tera, Giga, Mega.

   Therefore, the symbol for kilogram is always kg, never "Kg" or "KG." For Pascal the symbol is always Pa, not "pa" or "PA."

3. The symbols are never pluralized. The symbol for both kilogram and kilograms is always kg, never "kgs." Likewise, 32 square meters is written as 32 m², not "32 ms²."
4. Exponents are used to express powers. Therefore, 32 square centimeters is written 32 cm², not "32 sq. cm."
5. If the symbol is composed of two letters there is no space between them — they form a single symbol. Millimeter is written mm, not "m m"; and mm² means (mm)², not m(m)².

In the lists below, the "soft" metric equivalents are given in parentheses after the "hard" (exact) equivalents.

Length

The basic unit of the metric system is the *meter*. In establishing the length of the meter the developers wanted to use a basic fact of nature. After several changes the meter was finally defined as 1,650,763.73 wavelengths of the orange-red spectographic line of krypton 86. This definition is based on a physical condition that will always be the same, which makes it a true standard. From the basic unit of the meter the system of lengths is simply expanded in either direction in multiples of tens, which eliminates confusing fractions because all lengths can be expressed by using decimals.

millimeter	mm	0.001	m	1 in. = 25.400 mm (25)
centimeter	cm	0.01	m	
meter	m	1.00	m	1 ft. = 0.3048 m (0.30)
kilometer	km	1,000.00	m	

LENGTHS FREQUENTLY USED IN THIS MANUAL

Inches	Millimeters	
0.75	19.1	(20)
1.00	25.4	(25)
1.375	35.0	(35)
1.5	38.1	(38)
1.75	44.5	(45)
2.0	50.8	(50)
2.5	63.5	(65)
3.0	76.2	(75)
4.0	101.6	(100)
4.5	114.3	(115)
5.0	127.0	(125)
6.0	152.4	(150)

Feet	Meters	
50	15.24	(15)
100	30.48	(30)
200	60.96	(60)

As you will notice, the word *meter* is used in developing all other units of length. The prefixes used are either Greek or Latin words or roots. These same prefixes are used in describing other measurements — again, a simplification. Since metrics is in powers of ten it is extremely simple to make other conversions. For instance, how many millimeters are equal to an inch? Knowing that 2.54 cm equals 1 inch, and knowing that 10 mm equals 1 cm, it is simple to deduce that 1 inch equals 25.4 mm.

Metric Squares

The method for devising square measures is simple. For example, begin with a square land plot measuring one meter per side. This is a square meter (m^2). A square this size is also equal to one that measures 10 dm (since 10 dm = 1 m) and therefore is equal to 10 dm × 10 dm = 100 square decimeters. We can then say that 1 m^2 equals 100 dm^2.

Volume

To measure volume, or capacity, develop a cube 1 meter by 1 meter by 1 meter. Its volume is 1 cubic meter (m^3). This cube is the same size as a cube measuring 10 dm to a side, and therefore has a volume of 10 dm × 10 dm × 10 dm = 1,000 dm^3. Consequently, there are 1,000 dm^3 in 1 m^3. A cubic metric unit is always 1,000 times larger than the next smallest one.

The basic unit of measure for volume is the liter. The liter is defined as 1 kilogram of pure water at 4°C. The liter is also equal in volume to one cubic decimeter (1 dm^3); hence, the interrelationship of measurement of volume and length. To expand the metric volume, all that is needed is to expand the liter by multiples of 10 in either direction and apply the normal prefixes.

milliliter	ml	0.001	L	
centiliter	cl	0.01	L	
deciliter	dl	0.1	L	
liter	l	1.0	L	1 U.S. gallon = 3.785 L (3.75)
kiloliter	kl	1,000.00	L	1 U.S. quart = 0.946 L (0.95)

Temperature

The metric temperature scale is known as the Celsius scale, named after the person who invented it. The term *centigrade,* although technically correct, is not used because of the possible confusion it could cause in some countries where "centigrade" is used to denote a portion of a right angle.

In the Celsius system the freezing point of water is 0° and the boiling point of water is 100°. The difference is 100°, which fits well with the decimal notation of the metric system.

To convert Fahrenheit to Celsius, or vice versa, either of the formulas below may be used.

$$°C = 5/9(°F - 32) \quad \text{or} \quad 0.556(°F - 32)$$

$$°F = 9/5(°C + 32) \quad \text{or} \quad °C(1.8) + 32$$

Pressure

The SI unit used to measure pressure is the Pascal.

Some common pressure measurements used in the fire service are given below with their Pascal equivalents.

1 pound per square inch equals 68.94.757 Pa;

1 psi = 6.895 kPa.

50 psi	344.7 kPa	(350)
80 psi	551.5 kPa	(550)
100 psi	689.4 kPa	(700)

Friction loss per 100 feet of 2½-inch hose

100 gpm	3 psi	20.68 kPa	(20)
200 gpm	10 psi	68.94 kPa	(70)
300 gpm	21 psi	144.77 kPa	(150)
400 gpm	36 psi	248.15 kPa	(250)
500 gpm	55 psi	379.17 kPa	(375)

Velocity (Rate of Flow)

The SI uses liters per second to indicate rate of flow of water. For convenience, the lists below use liters per minute. Note that the symbol is L/min, to prevent the confusion that might result with l/m.

1 gpm = 3.785 L/min (3.75)

Pumper ratings	750 gpm	2,838.6 L/min (2,800)
	1,000 gpm	3,784.8 L/min (3,800)
	1,250 gpm	4,731.0 L/min (4,700)
	1,500 gpm	5,677.2 L/min (5,700)
	2,000 gpm	7,569.6 L/min (7,500)
Hand-lines	20 gpm	75.6 L/min (75)
	30 gpm	113.4 L/min (110)
	100 gpm	378.6 L/min (380)
	120 gpm	454.2 L/min (450)
	200 gpm	757.2 L/min (750)
	250 gpm	946.2 L/min (950)
Head	0.434 psi/foot	9.81 kPa/m(10)

4

THE APPLICATION OF FIRE STREAMS

NFPA STANDARD 1001

Firefighter II

4–8 Fire Streams.

4–8.1 The firefighter, given fire situations, for each situation shall:

(a) Identify the phase of burning

(b) Select the proper nozzle and hose size

4–8.3 The firefighter, given five fire ground situations, shall select and identify the proper adaptors or appliances.

4–8.4 The firefighter shall identify several precautions to be followed while advancing hose lines to a fire.

4–8.8 The firefighter shall identify three observable results that are obtained when the proper application of a fire stream is accomplished.

SIZE-UP

Size-up is first thing to do

When a company arrives at a fire, the first action of the person in charge is to do a size-up. The size-up is an evaluation of the fire in terms of its location, magnitude, fuel, phase, probable avenues of extension (both internal and external) and probable increase in magnitude. Along with these considerations are others, including life safety, weather, wind direction and speed, ventilation needs, available water supply, and the immediate and potential availability of apparatus, equipment and manpower. The size-up, in short, is an evaluation necessary to fulfill the ultimate objective of fire fighting: locate, confine and extinguish the fire. Obviously, the selection of the proper fire stream or streams is essential for such fulfillment.

Some of this selection should be based upon knowledge gained by pre-fire planning. Pre-fire planning includes not only information concerning water supply, building construction, fuel load, occupancy and probable avenues of attack, but also the standard operating procedures used by the fire department. Such standard operating procedures include the use of preconnected lines with nozzles attached, "quick-attack" operations, location of first-in apparatus, etc. Tangentially, pre-fire planning includes the training and knowledge of the firefighters. They should know the capacities of all nozzles and master stream appliances carried on their apparatus. (For complete pre-fire planning information see IFSTA 302, **Fire Department Facilities, Planning and Procedures.)**

FIRE FLOW: VOLUME AND RATE OF APPLICATION

The determination of adequate fire flow (the amount of water used for extinguishment) is arrived at by taking the size-up information into account. Enough water must penetrate to the area of the fire to absorb more heat than is being generated (remember that water's greatest extinguishing attribute is in cooling, rather than smothering). When the heat absorption is adequate, the main body of the fire should be controlled within 30 seconds.

Adequate fire flow depends on conditions

An adequate fire flow is, of course, variable depending on conditions, but many departments use rules of thumb that are accurate enough for the person in charge to make decisions concerning stream size and type, calls for additional alarms, etc. One of the most frequently used rules of thumb for enclosed, fully involved areas is that a fog nozzle discharging one gallon of water per 100 cubic feet of space (3.8 L/m^3) should produce enough steam to extinguish a fire within 30 seconds.

This is a relatively valid rule, but existing conditions may demand some revision of it at the fire scene. Of the number of variables that might have to be taken into account, one of the most obvious is that water distribution cannot always be even and efficient. For this reason, 3 or 4 gpm per 100 cubic feet (11.4 or 15.2

L/283 m³) is considered to be adequate by many departments. Obviously, when a fire department makes a rule-of-thumb fire-flow determination, consideration must be given to available and potential manpower and the capacities of apparatus and equipment.

When the fire is knocked down the total application should be stopped. Subsequent, short applications of straight streams (to avoid upsetting steam blankets and air currents) can be used on hot spots, etc.

The capacities of attack lines, nozzles and master stream appliances must be taken into consideration when fire-flow determinations are made. Two of the most common mistakes made is in providing too few hand-held streams of adequate capacity and in not providing master streams when they are needed. Also, there should be no hesitation in laying adequate hoselines to back up a fire attack even if they may not be needed immediately.

$$\text{gpm} = \frac{\text{ft}^3}{100}$$

As was mentioned above, water should be applied for as short a time as possible, to reduce water damage and prevent building collapse. Application should be stopped whenever flames are no longer visible or cannot be reached by the stream. The weight of water is an important consideration when large handlines or master streams are used. For example, one 600 gpm (2,275 L/min) master stream discharges approximately 5,000 pounds (2,268 kg) of water every minute. This added weight can result in structural collapse, especially if runoff is slow.

POINTS OF APPLICATION

The strategy of what is called the three-pronged attack is necessary to control and extinguish fires of any magnitude. The targets of the attack are exposures, avenues of fire spread and the seat of the fire. All three positions can be covered with only one or two handlines, if the fire is small. Existing conditions, of course, determine the number. types and volumes of fire streams that will be effective.

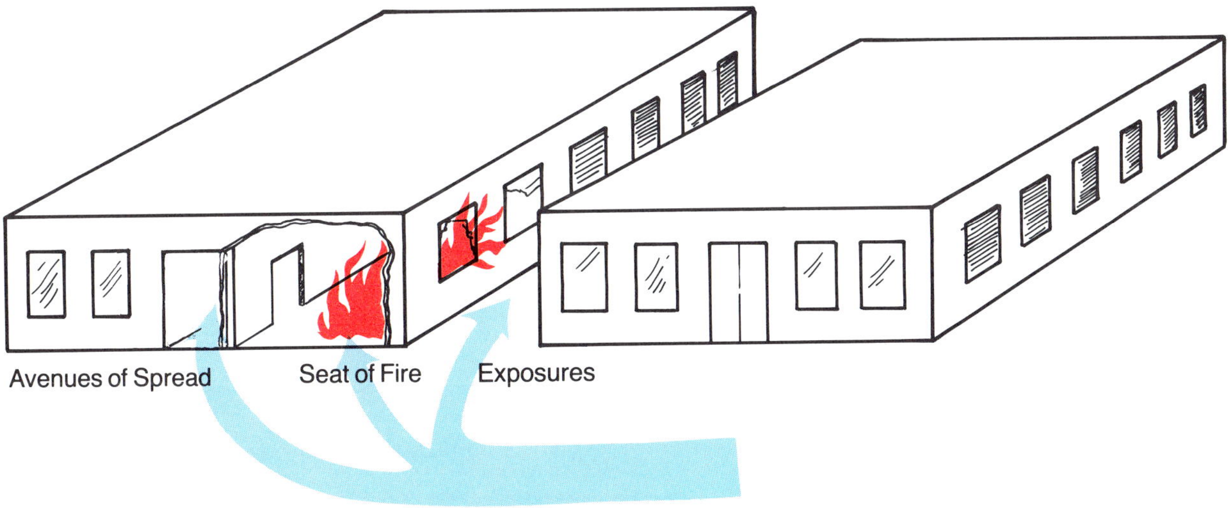

Figure 4.1 Three-pronged attack.

Exposures

Exposures must be protected

Exposures are those things, not directly attached to the involved structure, that could be consumed or damaged by a communicated fire or that could help to extend the fire. Exposures include nearby structures, combustible material, flammable liquid or gas storage, motor vehicles and ground cover (fields, woods, etc.). All windows, doors and other openings at the exposed side of a building should be protected with adequate fire streams.

The heat-absorbing capacity of water is of great aid in protecting exposures, especially if the water is in the form of a sheet, rather than as droplets or spray. Proof of this was found in an Underwriters Laboratories experiment. The temperature of a vertical steel plate was recorded while it was exposed to fire. A thin sheet, or film, of water was then allowed to flow continuously over the metal surface. It was estimated that less than 1 percent of the heat energy liberated by the fire during the first five minutes of the test actually passed the water-film barrier. The test further showed that only 5 percent of the heat stoppage was the result of absorption — 95 percent was reflected by the water surface.

This experiment indicates that a film of water is considerably more effective in absorbing and reflecting radiated heat from a fire than is water broken into droplets, such as with a water curtain. The best exposure protection, then, is not a water curtain, but water running over the surface of the exposure. This does not mean that a water curtain is worthless; only that if there is a choice, a film of water is better protection.

Avenues of Fire Spread

Fire can spread throughout a burning building by one or more of four methods, commonly referred to as conduction, radiation, convection and direct flame contact. These were discussed in Chapter 1.

Fire spreads throughout a building by a horizontal, vertical and downward extension process. Places that are vulnerable to fire spread are false ceilings, attics, cocklofts, corridors, vertical openings, horizontal openings and spaces between floors and ceilings and between walls. These areas and places should be seriously considered when a fire stream is selected. Some of the ways by which this extension is possible are listed below.

HORIZONTAL EXTENSION

- Through wall openings by direct flame contact or by convected air currents.
- Through open space by radiated heat or by convected air currents.

- Through corridors, halls or passageways by convected air currents, radiation and direct flame contact.
- In all directions by explosion or flash burning of fire gases, flammable vapors or dust.
- Through walls and interior partitions by direct flame contact.
- Through walls by conduction of heat through beams, pipes, ducts or other objects that extend through walls.

Avenues of spread are many

VERTICAL EXTENSION

- Through stairwells, elevators and shafts by direct flame contact or by convected air currents.
- Through partitions and walls and upward between the walls by direct flame contact and convected air currents.
- Through windows or other exterior openings by convection or direct flame contact into upper floors.
- Through ceilings and floors by conduction of heat through beams, pipes, or other objects that extend from floor to floor.
- Through ceilings and floors by direct flame contact.

DOWNWARD EXTENSION

- Through floors and ceilings by direct flame contact.
- Through floor and ceiling openings where sparks and burning material fall through beams, pipes and other objects that extend from floor to floor.

Using streams at avenues of fire spread means to get between the fire and the unburned parts of the building. By doing so and effectively using the fire streams the fire can be confined and extinguished.

Seat of the Fire

The seat of a fire is sometimes quite visible and easily identified, but at other times its location may be completely concealed and difficult to determine. The seat of the fire will usually be located where the most intense burning is present, unless the fire has extended to more highly flammable materials or an explosion has caused a rapid extension of the fire. The purpose of locating the seat and intense burning is to initiate an immediate, concentrated attack with fire streams at that point.

Locating seat of fire very important

Enclosed fires sometimes present a situation where the seat can be attacked before fire has had a chance to extend to other areas. In this case, an attack at avenues of fire spread and at exposures may not be necessary although hoselines and fire streams should in most instances be provided at these two positions. Existing conditions sometimes cause fires to spread so rapidly that it is not only desirable but essential to advance fire streams to exposures and avenues of spread before the seat can be reached.

When an attack is made at the seat of the fire the rate of flow of the attack lines must be sufficient to reduce accumulated heat and heat production, as was discussed earlier in this chapter.

CONFINING AND EXTINGUISHING FIRES

Confining a fire is accomplished by the well-planned use of fire streams, other extinguishing agents and ventilation. Particular attention must be given to the avenues and potential methods of fire spread. Charged lines should be positioned and manned as soon as possible to protect internal and external exposures (for example, at the top of stairways and similar vertical channels, as well as horizontal channels). In short, confining a fire means to block its extension, which calls for an aggressive, interior attack.

An aspect of planning for confining a fire, sometimes forgotten with disastrous consequences, is that the size-up must be thorough even if it takes a few minutes before the attack is begun. An ill-considered attack can by itself extend the fire, making extinguishment much more difficult, increasing the potential hazards to firefighters. An example of this is the common mistake shown in Figure 4.2. The firefighters in this illustration have forgotten the cardinal rule "Fight from the *unburned* portion of the building" and are actually spreading the fire, rather than confining and extinguishing it.

Figure 4.2 Improper use of a fog nozzle can spread fire through a building. This can be prevented by adequate size-up, fighting the fire from the unburned part of the building, and covering avenues of spread.

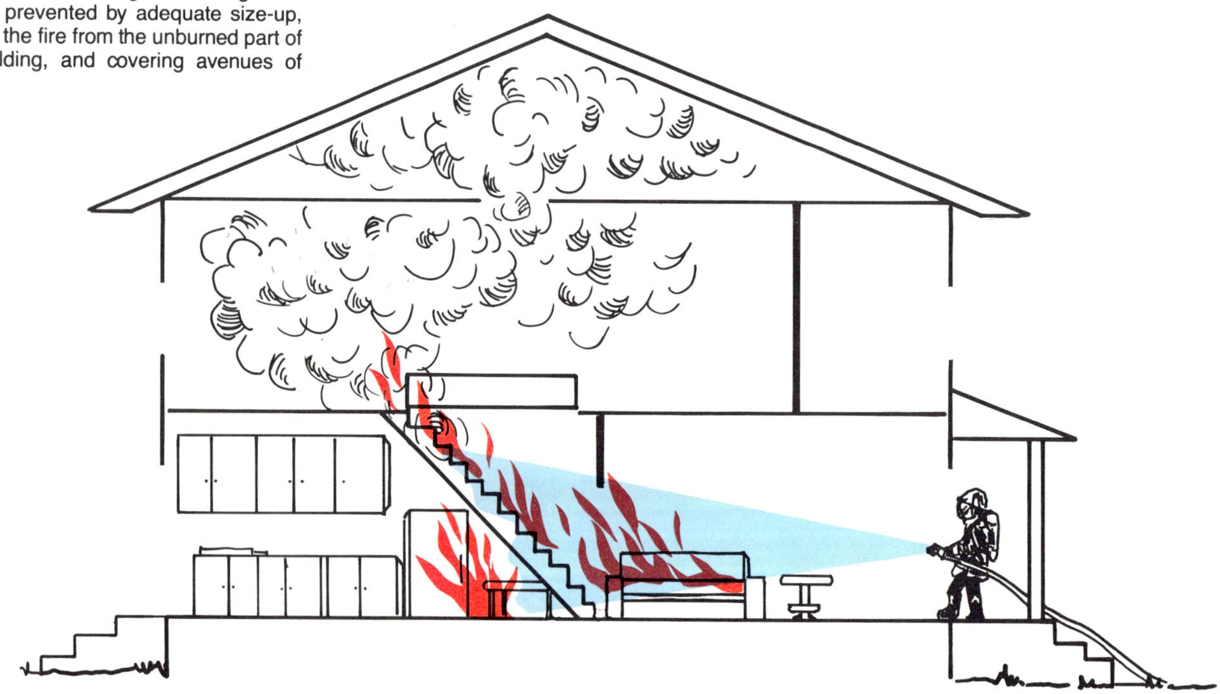

Firefighters are usually able to attain a high percentage of efficiency when extinguishing small or medium fires because the extinguishing agent is usually adequate and properly applied. Fires of a greater magnitude, however, may require more time and a concentration of water application. The first step toward final extinguishment is to ventilate and quench the main body of the fire. This action will immediately reduce heat generation to a point where firefighters can either enter the burning area or advance closer to the fire. Spot fires will generally warrant a second step toward final extinguishment: a closer approach and a direct application of water to burning material. A third step toward final extinguishment is to locate and extinguish hidden fires. Hidden or deep-seated fires are quite different from the so-called spot or area fires. Locating and extinguishing hidden fires is considered to be an overhaul operation. Information concerning how to locate and extinguish hidden fires can be obtained from IFSTA 104, **Salvage and Overhaul Practices**.

Three steps to extinguish fire

ADVANCING THE NOZZLE: PRECAUTIONS

Before a hoseline and nozzle are advanced to fight a fire there are several precautions that must be taken if the firefighters are going to survive and the fire is going to be controlled and extinguished. One of the first of these precautions is that of ventilation. If there is any possibility of a backdraft, the structure must be ventilated from the top before firefighters attempt to enter the structure. Other precautions involve personal safety. Many firefighters forget, under the stress of excitement, that their own safety is a prerequisite for the safety of the people and property being protected by the fire department. Some of these essential personal safety precautions are listed below:

Personal safety must be remembered

- Wear proper protective clothing.
- Positive-pressure, self-contained protective breathing equipment should be worn to provide protection against toxic gases that may be encountered.
- When advancing through a smoke-filled area, a firefighter should feel the way. This will avoid falling through burned floor areas, elevator shafts, pits or stairways.
- Feeling the walls or contents with the hands will help to guide a firefighter around obstacles when advancing through smoke-filled areas.
- When entrance is made, firefighters should advance in pairs or groups and never without a charged hoseline.
- If it is necessary to retreat, close the nozzle and follow the hoseline to the outside of the building.

- The value of water fog in dissolving and settling gases of combustion and in expelling smoke should not be ignored or taken lightly. Water fog should be used as a safety precaution when firefighters are entering a highly heated area.

- Instruments to detect flammable and poisonous gases should be used whenever feasible.

- Where the presence of radioactive materials is suspected, proper precaution and protection should be used. See IFSTA 209, **Firefighter Safety**; NFPA, *Handling Radiation Emergencies;* and IAFC, *Nuclear Hazard Management for the Fire Service.*

Be alert for changes in conditions

Many hazardous situations will present themselves in actual practices and they must be met as they arise. Many of these hazardous situations cannot be avoided, but proper protective measures can be taken.

SMALL STREAMS

A small stream is a low-volume stream, either solid or fog, discharging less than 40 gallons per minute (150 L/min). These streams include booster lines and some 1½-inch (38 mm) hose streams. Applying such streams usually involves only one person and their use is generally limited to the first phase of fire development. Table 4.1, "Analysis of the Extinguishment of a Confined Structural Fire," shows that the first phase of a beginning fire requires only a small stream for extinguishment. It further points out that extinguishment is accomplished by direct application at the base of the fire. It should be kept in mind that the limits of the small stream are reached when heat is being generated more rapidly than the rate-of-heat absorption of the small-volume stream.

SOLID STREAMS

Solid streams sometimes best

The fire service has done an almost complete turnaround in the last few years, from considering fog streams limited in usefulness to considering them a cure-all. This impression, of course, is a fallacy — there is no universal method or agent for extinguishing fires — but its prevalence has shoved the solid stream into the background.

Although fog streams have been improved, the reach and penetration at the straight stream setting is not as great as that of comparable solid streams. The straight stream has less density and is more susceptible to disruption from wind and from friction with the air. Solid streams, then, should be used whenever reach and penetration are needed.

FOG STREAMS

The use of water fog in fire extinguishment requires a special technique of operation. The mere fact that firefighters have a good fog nozzle that supplies a protective curtain does not enable them to advance helter-skelter into a heavily charged area and expect to do an effective ventilation and extinguishing job. When water fog is used for ventilation and extinguishing purposes, the degree of effectiveness depends upon how, where and when the fog stream is applied. As the fog stream is projected, it should be "worked"; that is, it should be directed to cover the greatest possible area. Tests have shown that a circular motion of the nozzle is usually most satisfactory.

Confined Structural Fires

The conditions that may be found in a confined structural fire and the phases of its development have been defined in Chapter 2. These same conditions have again been outlined by Table 4.1.

TABLE 4.1
Analysis of the Extinguishment of a Confined Structural Fire

Phase	Conditions	Requirements to Extinguish	Method
First	Beginning fire (incipient). 20% oxygen content of atmosphere. Accelerating heat generation according to development of fire. Considerable smoke in reverse proportion to heat generation. Destruction limited to immediate fire area.	Entrance and direct application of extinguishing agent. Personal protection if needed.	Make entrance. Locate fire. Apply extinguishing agent at base of fire. Protect self with protective breathing equipment. Ventilate at natural openings if needed.
Second	Flame production. Major involvement. Entrance difficult because of heat. Oxygen reduction 21 - 15% or less. Rapid heat generation with highest temperature at ceiling. High convection of heated gases and air. Limited smoke production, increasing as flame decreases. Increasing atmospheric pressure within space. Rapid destruction.	Reduction of room temperature to 300°F or less. Entrance and direct application of extinguishing agent to remaining fire. Personal protection if needed.	1. Locate area of major involvement. 2. Determine point or points at which application is to be made. 3. Make opening, if needed, small as possible for application of water fog. 4. Assume position so that nozzleman will not be involved with hot smoke and condensing steam. 5. Apply fog into upper level to meet space requirements. -If applied at upper level use short-reach fog. -If applied at lower level use long-reach fog and direct into upper level.
Third	Smoldering fire. Oxygen below 15%. Convection ceased and stratification of temperatures, highest at upper level. General decreasing of temperature due to lowering of heat generation and heat losses. Intense smoke production. Serious accumulation of carbon monoxide gas. Chance for atmospheric breathing. Possibility of smoke explosion if air is admitted.	Reduction of temperature. Displacement of smoke and gases with steam to make space livable and eliminate chance for explosion. Entrance and direct application of extinguishing agent. Personal protection if needed.	The order of expulsion due to displacement will be: 1st. Smoke 2nd. Smoke and condensing steam 3rd. Condensing steam 6. Continue application of fog without interruption until decrease of condensing steam is noticed. 7. Enter and extinguish remaining fire and spot fires. 8. Protect self if necessary with breathing equipment and heavy clothing.

Some of the important factors for applying fog streams include:

- If there is any possibility of a backdraft, be sure top ventilation has been accomplished before applying the stream. Sometimes the entrained air of a fog stream can be enough extra to cause a backdraft even after a door or window has been opened.

Project fog stream into hottest space

- The fog stream should be projected into the hottest space, which may be the upper levels if the fire has burned for a considerable duration (taking care to not upset thermal layers).
- If a wide-angle fog pattern will cover the area, it should be projected upward into the heated area and worked in a circular motion for complete coverage.
- If the area is so large that it cannot be reached by the wide-angle fog pattern, a narrow-angle fog pattern should be used to carry the droplets into the hottest space and the nozzle worked back and forth for complete coverage.
- The number of fog streams to use should be determined by the cubic feet of the area involved and the volume of the nozzles (see "Fire Flow: Volume and Rate of Application").
- Enough water fog must be applied to absorb the heat faster than it is being generated.
- One or more extra lines should be supplied for auxiliary use. They may be used to extinguish the remaining fire.

Three indications of extinguishment

- The officer in charge should watch for displacement indications: first, smoke; second, smoke and condensed steam; third, condensed steam.
- A fog stream should be shut off as soon as smoke and gas have been displaced by condensed steam.
- If the atmosphere is still too hot, a second burst of water fog may be required.
- All spot fires may be extinguished by direct application.
- When the application of water fog is made in a building, the firefighters should keep well out of line with openings to protect themselves from the expulsion, through the openings, of heated gases and steam.

Elevated Fog Streams

Elevated fog streams are usually used to cut down sparks and flying brands from a building or to reduce the thermal column of heat over a building. Established ventilation can be disrupted by an improperly directed elevated stream. This disruption can also make the building interior untenable for firefighters. When elevated streams are projected downward and through a ventilation opening, or even when they are used to reduce the thermal column to a point where ventilation is hindered, they either destroy or upset the orderly movement of fire gases from the building. Figure 4.3 illustrates how improperly directed elevated streams can upset ventilation and advancement operations.

Elevated streams operating closely above ventilation openings should be projected horizontally above the opening. In this position they will help to cool the thermal column and extinguish sparks. Figure 4.4 shows how properly directed elevated streams can help the ventilation process and aid fire fighting.

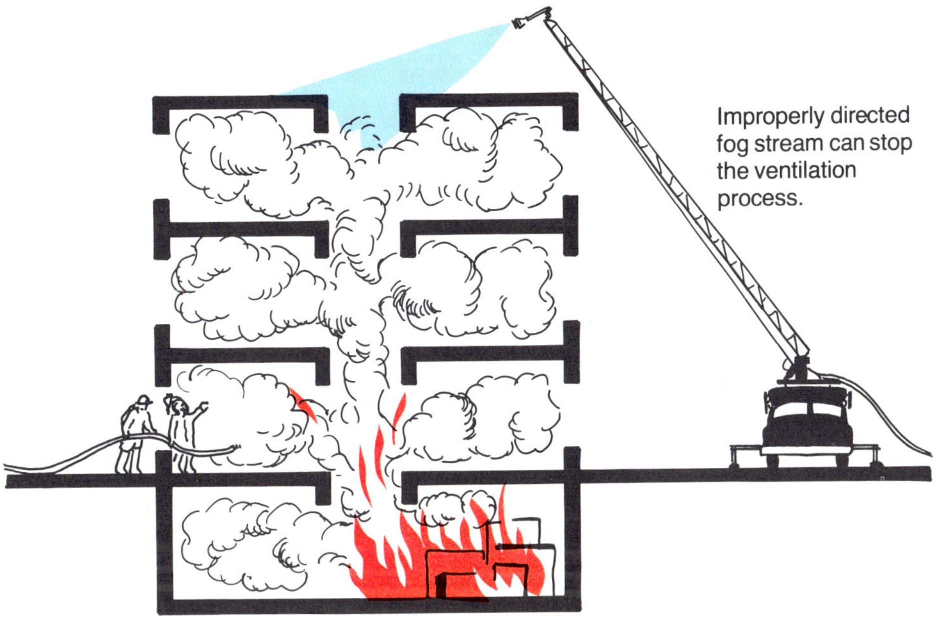

Figure 4.3 Improper positioning of elevated fog nozzle works against ventilation and endangers firefighters in building.

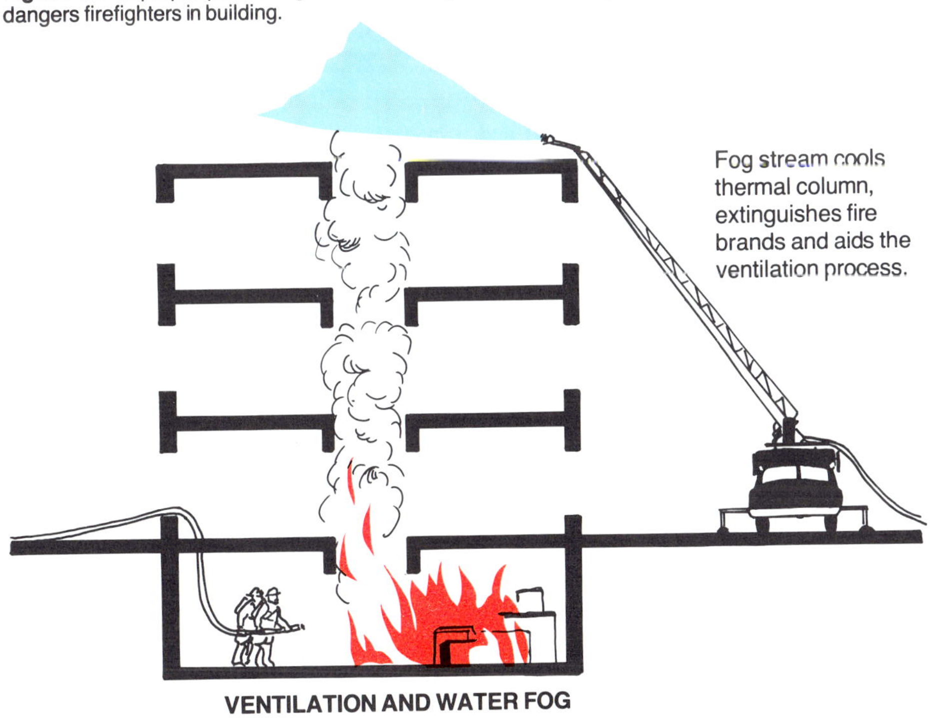

Figure 4.4 Proper positioning of elevated fog nozzle does not hinder ventilation; helps to cool thermal column and control sparks and brands.

Multiple-Line Attack

Occasionally fire companies must advance two or more handlines with fog nozzles for a concentrated attack. The fog streams have a dual purpose: to push away or diminish the intensity of fire in front of the advancing firefighters and to protect the nozzlemen, rescuers or other persons involved in the advance. The streams must be discharged in the most efficient patterns for the particular fire situation. This technique might be necessary for an aircraft crash fire or rescue, a flammable liquid spill fire, or perhaps for protecting a person trying to close a valve during fire conditions.

Teamwork essential for multiple-line attack

Teamwork is essential when advancing two or more lines together. During a flammable liquid fire, for example, a lack of teamwork can cause conditions favorable for a backflash, which can envelop firefighters instantly. Every firefighter involved in a multiple-line fire attack has a responsible role. Each pump operator must be alert and supply the correct pressures to produce the most effective nozzle discharge. Each nozzleman must consider the personal mission and safety as well as that of every other firefighter participating in the attack. The nozzlemen must rely on training and experience to direct the fog stream in an effective pattern to coincide with the stream directed by other nozzle. Each person handling a line must be sure that no motion is wasted and that the hose is moved so that the nozzleman's work can be done with a minimum of effort. The protective effect of the stream patterns must intermingle. When two teams are advancing together, the volume discharged from each nozzle should coincide, and the fog patterns should be similar. The officer supervising the advance must exercise good judgment of the potential effect of the fire attack and the relative dangers to the firefighters involved.

Interlock fog patterns

When two or more lines are advanced in a concentrated attack, the fog patterns should be interlocked to produce a solid mass of water fog. All firefighters should be on the inside of the hoselines, to receive maximum protection. There are many things that can happen during an operation of this type. For example, one of the nozzlemen could fall, a hoseline could burst or a gust of wind could disperse the fog stream. It should be standard operating procedure to always have one or more additional protection lines following behind or beside the advancing lines.

A fire situation may make it necessary for an officer to advance with the nozzlemen and give commands. Commands should be given audibly and reinforced with hand signals, by pressure or pulling on the firefighter's shoulders. Drill with charged lines during training to insure complete understanding before hot fire drills. When several attack lines are used, all firefighters should be alert for changes in direction or stream patterns.

Hoselines cannot be properly advanced unless sufficient personnel are available to move the lines. This factor is not only important in the advancement but also in backing out. Nozzles and hoselines should never be dropped in an attempt to run back. When fighting flammable liquid fires, the backout must be slow. For most situations, all fog patterns should stay the same during the advance or retreat because the adjustment of only one fog pattern may break the protective curtain.

MASTER STREAMS

Master streams are sometimes referred to as heavy streams. Although situations that require these streams are relatively rare, they present a great possibility of disaster. Because of this possibility it is vital that master streams be placed in service with minimum delay when needed. The danger involved and the infrequent use of master streams make it necessary for firefighters to repeatedly practice their production and have a clear understanding of all factors involved.

The principles upon which master streams are produced are essentially the same as for hand streams, because the same physical laws apply. To obtain larger volumes and greater reach, higher nozzle pressures must be used on solid master streams. Fixed nozzles are not limited by nozzle reaction or by requirements of maneuverability. The nozzle pressures required to produce effective master streams may vary under different conditions because of the wide range of master stream devices used to produce these streams. For average use, however, the generally recognized nozzle pressure is 80 psi (550 kPa) for master solid streams and 100 psi (700 kPa) for master fog streams.

An aerial ladder pipe or elevating platform may be used to direct water in the form of a master stream. This method of applying master streams may offer the advantage of being able to get above or on the same level with the fire (Figure 4.5). The volume from older master stream devices may be limited because of the small throat found in the piping of some ladder pipes and deck guns.

Figure 4.5 Ladder pipe in operation. Note that stream enters windows horizontally. A stream entering above the horizontal would be ineffective.

In the tests conducted by an Exploratory Committee on the Application of Water (IFSTA/FDIC), the effect of master fog streams in the heated area above a fire was clearly apparent. The results of these tests indicated that the proper application of master fog streams may not extinguish the fire, but that they definitely permit firefighters with hand streams to approach much closer to the burning material. The use of master streams may be warranted on lumber yard fires, gasoline transport fires, dock and wharf fires, and other fires that may cover large areas.

The effectiveness of master streams depends upon penetration and deflection of the stream. The location of a master stream device determines the height at which the stream can maintain adequate penetration and deflection. For example, a master stream device that is placed on the ground cannot penetrate floors that can be reached by master streams directed from fire apparatus, ladder pipes or elevating platforms. To attain the maximum height and still penetrate into the upper floors, master streams are usually located as far as is practically possible from a burning building. The penetration of master streams at different angles as they are delivered from an apparatus-mounted master stream appliance is illustrated in Figure 4.6.

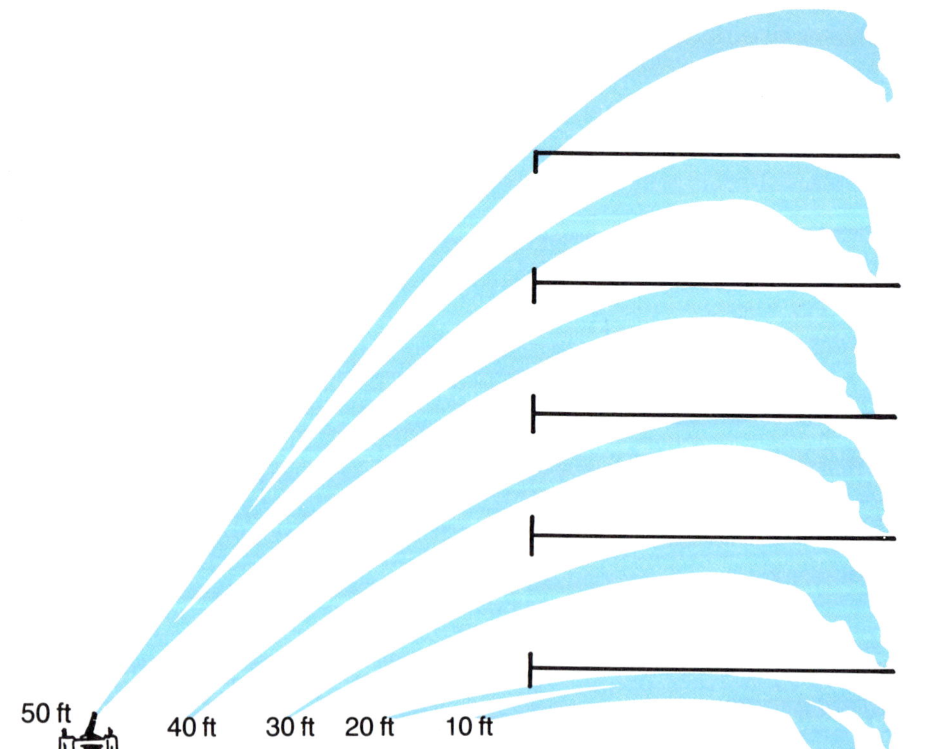

Figure 4.6 Penetration of an apparatus-mounted master stream. A fire stream is not a solid, and will not act like a billiard ball when deflected by a plane surface. Rather, the water will flow for a distance before dropping (dependent on velocity, size, type, and angle of meeting the surface).

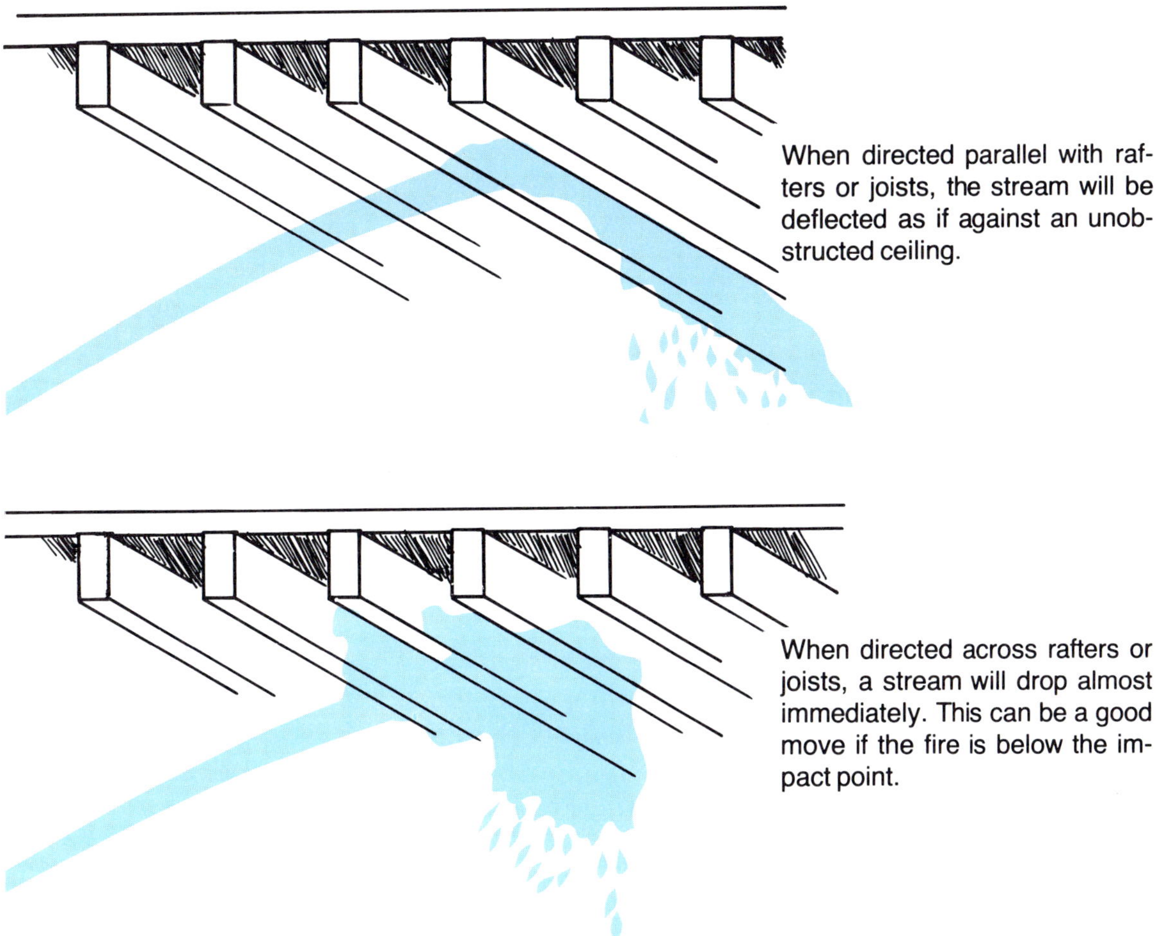

Figure 4.6a Foreknowledge of a building's construction is important for applying fire streams.

BROKEN STREAMS

Broken streams produced by broken stream nozzles (see "Broken Stream Nozzles", Chapter 2) are especially useful for hidden fires in walls, ceilings and partitions, and for attic and basement fires. The most effective nozzle to use will be determined by the magnitude of the fire, its location and fuel. More than one hole for inserting the nozzle may be necessary and somtimes more than one nozzle will be needed to attain adequate flow and application. Remember the rule of thumb for rate of flow and application: If the rate of flow is adequate but the fire is not knocked down within one minute, the seat of the fire is not being reached.

Unseen obstructions will inhibit the effectiveness of a broken stream nozzle used through a hole. These obstructions include studs, firestops, piled stock in basements and storage in attics. Because of this, a clear area for the stream should not be taken for granted.

Figure 4.7 Basements — Lower nozzle to check for obstructions or this might happen.

Figure 4.8 Attics, Upper Floors — Push nozzle up or use pike pole to check for obstructions or this might happen.

When used for a basement fire, the nozzle should be lowered until it hits something solid, to check for obstructions. If the drop is far enough to indicate that there are no obstructions between the floor and ceiling of the basement, raise the nozzle a little higher than halfway and then apply the stream.

An attic fire may be attacked from beneath, through the ceiling, or from above, through the roof. When used through the ceiling, push the nozzle up, as far as is possible, to check for obstructions. When used through the roof, check for obstructions by lowering the nozzle, as for basement fires.

Broken streams created by the impingement of two or more solid streams can be useful as water curtains. They can also be used for outside fires with the same effect as master streams.

Figure 4.9 Impinging two solid streams (1- or 1⅛-inch tips) will create a broken stream of "heavy rain." This can be useful for outside fires, with the effect of a master stream.

5

SPECIAL INFORMATION

NFPA STANDARD 1002

3–4 Hydraulic Calculations.

3–4.3 The fire apparatus driver/operator shall identify the following terms that relate to the basic principles of fire service hydraulics . . .

(b) Capacity

(c) Displacement

There are special fields of information instrumental to the production of fire streams. The use of measuring devices and instruments, mathematics and fire stream tables are all relative to fire streams and their use.

DEVELOPING FLOW FORMULA

The gallons per minute delivered depends upon the size of the orifice and the rapidity, or velocity, of the flow. Thus, a 200 gpm (750 L/min) delivery means that 200 gallons (750 l) pass a given point each minute. Velocity is the speed of an object in a given direction. Direction and speed are both essential parts of velocity. As water is forced from an orifice it is carried forward by its momentum, usually referred to as forward velocity. Forward velocity is generally measured in feet per second. The discharge of water from any circular orifice is equal to the *area* of the outlets multiplied by the *velocity* of the flow. The formula is written

$$gpm = AV$$

where

gpm = Gallons per minute
A = Area in square inches
V = Velocity in feet per second

Since gpm is measured in gallons per *minute,* the area (A) times velocity (V) must be multiplied by 60 to convert feet per second to feet per minute; hence

$$gpm = 60AV$$

Also, since area (A) is stated in square inches and velocity (V), in feet per minute, the velocity must be multiplied by 12 to convert it to inches per minute; hence

$$gpm = 60A\,12V$$

The formula is now stated in inches per minute. When area (A), which is figured in square inches, is multiplied by velocity (V), which is now stated in inches, the result will be cubic inches. Since there are 231 cubic inches in a gallon, the result must be divided by 231; hence

$$gpm = 60\frac{A12V}{231}$$

By formula substitution:

$$A = \pi r^2 \qquad \text{or} \qquad A = \frac{\pi d^2}{4}$$

Therefore

$$gpm = 60\,\frac{\pi d^2 12V}{4(231)}$$

So far, all quantities have been accounted for except for velocity. The derivation of the formula for velocity is as follows.

From the law of falling bodies, the following formula is used as the basis for determining velocity:

$$S = \tfrac{1}{2}gt^2$$

where

S = Distance

g = Acceleration because of gravity (32 feet per second^2)

t = Time

A column of water 2.304 feet high exerts a pressure of one pound per square inch. Therefore, h can be substituted for S in the formula:

$$h = \tfrac{1}{2}gt^2 \qquad \text{or} \qquad h = \frac{gt^2}{2}$$

Velocity is equal to acceleration multiplied by time, which gives the following formula:

$$V = gt$$

Using this formula as a starting point, the formula for finding t is

$$t = \frac{V}{g}$$

By substitution, the formula $h = \tfrac{1}{2}gt^2$ becomes

$$h = \left(\frac{g}{2}\right)\left(\frac{V}{g}\right)^2$$

$$h = \left(\frac{g}{2}\right)\left(\frac{V}{g}\right)\left(\frac{V}{g}\right)$$

$$h = \frac{V^2}{2g}$$

This formula is then solved for V, which gives

$$V^2 = 2gh$$

$$V = \sqrt{2gh}$$

The value of g is 32, so the formula can be modified further:

$$V = \sqrt{2(32)h}$$

$$V = \sqrt{64h}$$

$$V = 8\sqrt{h}$$

Using the formula for the relationship of head in feet and head pressure, $h = 2.31P$, and substituting in the velocity formula:

$$V = 8\sqrt{2.31P}$$
$$V = 8(1.52)\sqrt{P}$$
$$V = 12.14\sqrt{P}$$

With this formula, the velocity (V) in the gpm formula can be accounted for:

$$gpm = 60\left(\frac{\pi\ d^2 12V}{4(231)}\right)$$

$$gpm = 60\left(\frac{\pi\ d^2 12(12.14\sqrt{P})}{4(231)}\right)$$

$$gpm = 60\left(\frac{3.1416d^2 145.68\sqrt{P}}{924}\right)$$

$$gpm = 60\left(\frac{457.668d^2\sqrt{P}}{924}\right)$$

$$gpm = 60\ (0.495d^2\sqrt{P})$$

$$gpm = 29.7d^2\sqrt{P}$$

The gpm flow of a hoseline can now be determined if the diameter of the nozzle and the nozzle pressure are known.

USING PITOT TUBE AND GAUGE

The pitot tube and gauge is used to find the velocity pressure at a discharge opening. The tube and gauge are connected to a handle made from pipe fittings. The handle serves as an air chamber to help keep the gauge indicator steady. By knowing the pressure at the nozzle and the size of the orifice, the gallons per minute discharge can be determined by using the formula $gpm = 29.7d^2\sqrt{P}$. This formula is for a smooth orifice. For other types of openings, such as hydrant butts, a factor should be used (see IFSTA 205, **Water Supplies for Fire Protection**).

The use of a pitot tube and gauge when taking a flow reading is not difficult, but certain rules must be followed if accurate readings are to be obtained. A good method of holding a pitot tube and gauge at a hydrant outlet or nozzle is illustrated in Figure 5.1. The pitot tube is grasped with the first two fingers and thumb of the left hand just behind the blade while the right hand holds the air chamber. The little finger of the left hand rests upon the hydrant outlet or nozzle tip to steady the instrument. Unless some

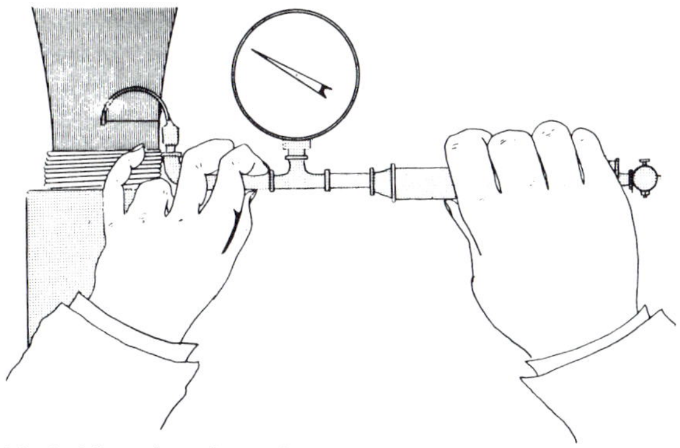

Figure 5.1 Method for holding pitot tube and gauge.

effort is made to steady the pitot tube while taking a reading, the movement of the water will make it difficult to get an accurate reading. The point or small opening in the pitot tube should be centered in the stream. The tube should be held away from the opening at approximately one-half the diameter of the opening. (For a 1½-inch [38 mm] opening, this distance would be ¾-inch [20 mm].) The pitot tube should be held at a right angle to the stream. The air chamber should be kept above the horizontal plane through the center of the stream for the further purpose of increasing the efficiency of the air chamber. After each test, the petcock at the end of the air chamber should be opened and all of the water should be drained from the assembly. This method is applicable only to 3-inch (75 mm) and smaller outlets. For larger openings, sample the stream for the highest reading.

PUMP CHART

Quite often, errors are made when making complex hydraulic calculations on the fireground. To reduce these errors a pump chart similar to the one shown as Table 5.1 may be developed. A

TABLE 5.1
Sample Pump Chart

			Length of Lay in Feet															
Nozzles	**GPM**	**NP**	**100**	**200**	**300**	**400**	**500**	**600**	**700**	**800**	**900**	**1000**	**1100**	**1200**	**1300**	**1400**	**1500**	**1600**
Booster	23	100		130														
Preconnect	60	100		130														
Wye	95	100	165	175	180	190	200	210	220	225	235	245						
2½" Fog	250	100	115	130	145	160	175	190	205	220	235	250						
1¾" Deluge	800	80	125	160	200	235	(2 - 2½" lines)											
1¾" Deluge	800	80	105	125	140	160	175	190	210	225	245	(3 - 2½" lines)						
1¾" Deluge	800	80	105	120	135	150	165	180	195	210	225	240	(2 - 3" lines)					
Relay 3"	250	20	25	30	40	45	50	55	60	70	75	80	85	90	100	105	110	115
Relay 3"	500	20	45	65	90	110	135	160	170	195	215	240						
Relay 3"	750	20	70	120	170	220												
Sprinkler		150 Maintain regardless of length lay																
Elevation Loss/Gain ½ psi per foot																		

chart of this type will allow the pump operator to accurately determine the correct pump discharge pressure for any given layout included in the chart.

To develop and use the pump chart, the column headings must be understood. The *Nozzle* heading should include only those nozzles and devices used by the department developing the chart. Also included are other layout descriptions, such as *Sprinkler* or *Relay 3-inch.* The *GPM* column indicates the flow being provided to that nozzle or layout. The *NP* column indicates the nozzle pressure being produced. The columns headed *100, 200,* etc., indicate the number of feet of hose being used to supply a given nozzle or layout.

To use the chart, locate in the *Nozzle* column the nozzle or layout being used. Follow that line across to the vertical column headed by the number of feet of hose in that layout. The figure in the block where the two columns intersect is the required pump discharge pressure.

The first step in developing a pump chart is to identify all the nozzles, devices and layouts used by the department and enter them in the *Nozzle* column. The gpm flowing and the nozzle pressure desired for each item should be entered in the appropriate columns. If, for instance, a deluge set might be supplied by either 2½-inch (65 mm) or 3-inch (75 mm) hose, a separate entry should be made for each layout in the *Nozzle* column, as indicated in the sample chart.

The second step is to calculate the required pump discharge pressures for each of the listed layouts, using the formulas and tables given earlier in this manual or by field testing. The following rules should be observed when making these calculations:

- Be certain to include friction loss in wyes, deluge sets, etc.
- For wye layouts, the length-of-layout numbers indicate the number of feet of hose between the pumper and the wye. (If used as a preassembled unit the 1½-inch hose leading from the wye is constant and therefore is not a factor in determining the length of the layout.)
- When a device may be supplied by a different number or size of hoselines, these should be indicated in the chart, as shown in the sample.
- Pump discharge pressure should be rounded off to the nearest five pounds.
- Pump discharge pressures should not be listed that exceed the test pressure used for the size of hose concerned.

- When calculating pump discharge pressures for relay operations, 20 psi (140 kPa) residual pressure should be provided at the intake of the pumper being supplied. This residual pressure may be indicated on the chart as a nozzle pressure.

The reliability of the chart will depend entirely on the accuracy of the calculations. Once developed, the chart should be placed in a clear, weather-proof cover and located on the pumper so it will be readily available to the pump operator.

CAPACITY AND DISPLACEMENT

The rated capacity of a fire pump is not necessarily the maximum capacity of the pump. The actual capacity of a centrifugal pump is limited by its design: the intake diameter, impeller eye diameter, outside diameter, width of the impeller, shape and number of vanes in the impeller, and design of the volute chamber. The capacity of a positive-displacement pump is limited by its displacement. This relationship is easiest seen by using the piston pump as an example (Figure 5.2).

Actual and rated capacity often different

The intake stroke of the piston pump creates a vacuum in the cylinder; atmospheric pressure forces water into the cylinder at the same time. The amount of water drawn into the pump (displacing the air) is the displacement.

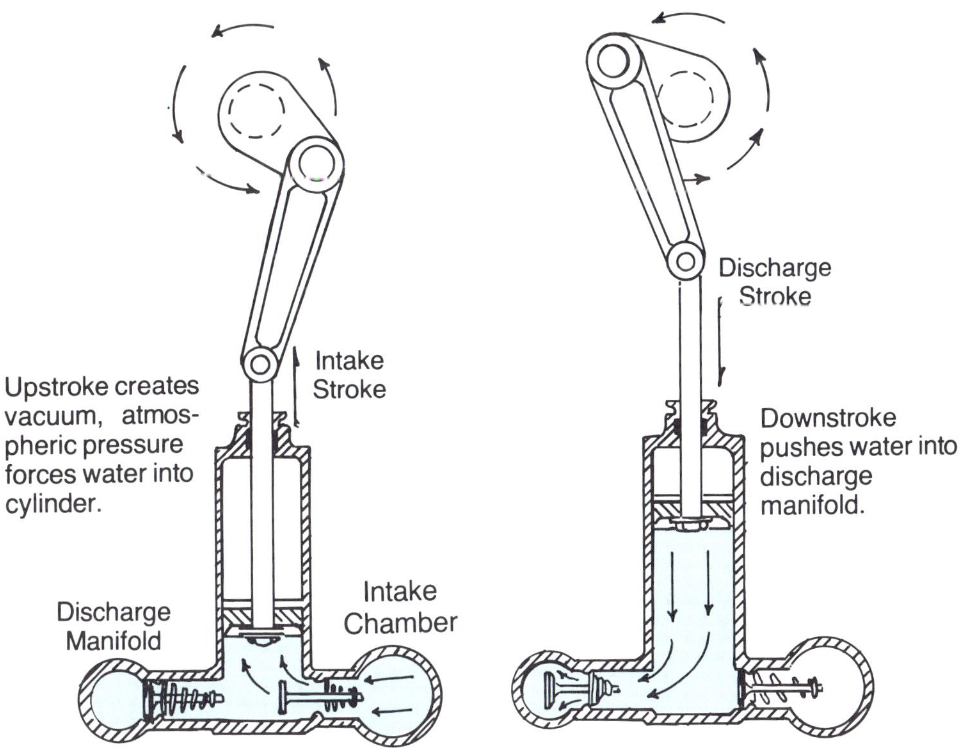

Adapted from Casey, *Fire Service Hydraulics*

Figure 5.2 How a piston pump works.

The formula for determining a piston pump's theoretical displacement in U.S. gallons is

$$D = \frac{ASn}{231}$$

where

D = Displacement in gallons per revolution
A = Cylinder area
S = Length of piston stroke
n = Number of cylinders
231 = A constant, the cubic inches in one U.S. gallon

Example

Using the formula for theoretical displacement, what is the displacement of this pump if it has three cylinders?

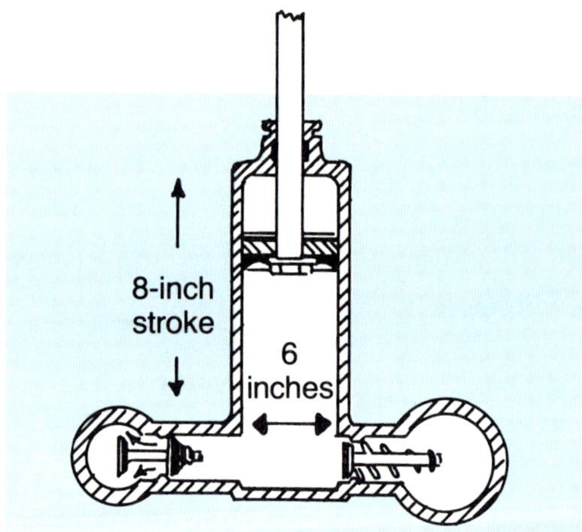

Figure 5.3

First, find the area of the cylinder:

$$A = \pi r^2$$
$$A = 3.1416(3^2)$$
$$A = 28.274$$

Then substitute the correct values in the equation:

$$D = \frac{28.274(8)(3)}{231}$$

$$D = \frac{678.57}{231}$$

D = 2.93 gallons per revolution

The displacement of a rotary pump depends on the design of the rotors. The shaded area of Figure 5.4 shows the displacement of a rotary pump.

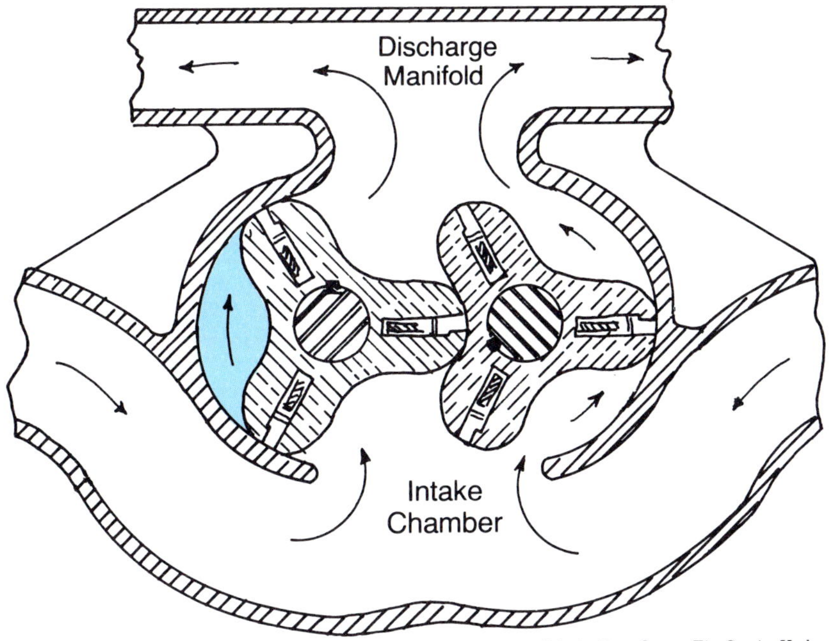

Adapted from Casey, *Fire Service Hydraulics*

Figure 5.4

Centrifugal pumps have largely replaced positive-displacement pumps as the workhorses of the fire service. They are more efficient and can use as much water as is supplied to them. The centrifugal pump uses incoming water pressure to its advantage. If, for instance, the incoming water pressure is 50 psi, the pump's net engine pressure (the total work the pump must do) is that much less. With a positive-displacement pump, the flow and pressure of incoming water makes no difference in discharge capacity. The centrifugal pump, unlike the positive-displacement pump, cannot create a vacuum, so it cannot draft water. A positive-displacement pump is used to prime the centrifugal pump.

APPENDIX

LIST OF EQUATIONS

(ENGLISH)	(SI)

Area of Circle

$A = \pi r^2$ or $A = \frac{\pi d^2}{4}$ — (SI) same

Capacity of a Cylindrical Tank

$gallons = d^2 6L$

where d = Diameter in feet
L = Length in feet
6 is a constant

$V = \pi r^2 L$ — (SI) $V = \pi r^2 L$

Velocity (GPM)

$gpm = 2{,}917 d^2 \sqrt{P}$

where d = Diameter in inches
P = Nozzle pressure

Waterflow

(English)

$Q = \frac{gpm}{100}$

(SI)

$Q = 0.06 d^2 \sqrt{NP}$

where d = Diameter (mm)
NP = Nozzle pressure (kPa)

Nozzle Reaction, Solid Streams

(English)

$NR = 1.5 d^2 NP$

where NR = Nozzle reaction (pounds)
d = Nozzle diameter (inches)
NP = Nozzle pressure (psi)
1.5 is a constant

(SI) [for either solid or fog]

$NR = \frac{Q\sqrt{NP}}{639}$

where NR = Nozzle reaction (kg)
Q = Flow (L/min)
NP = Nozzle pressure (kPa)
639 is a constant

Nozzle Reaction, Fog Streams

$NR = 0.0505 Q \sqrt{NP}$

where NR = Nozzle reaction
Q = Flow (gpm)
NP = Nozzle pressure (psi at *base* of nozzle)
0.0505 is a constant

(ENGLISH)	(SI)
Friction Loss, 2½-Inch Hose (less than 100 gpm) $FL = 2Q + 0.5Q$	$FL = Q^2 + Q$
Friction Loss, 2½-Inch Hose (equal to or greater than 100 gpm) $FL = 2Q^2 + Q$	
Friction Loss (other than 2½-inch hose) $FL = 2Q^2 + 0.5Q$ then multiply or divide FL by factor found in Table 3.2.	

SI EQUATIONS — FRICTION LOSS WHEN FLOW IS KNOWN

Hose Diameter	Equation
25 mm	$FL = 140Q^2$
40 mm	$FL = 13Q^2$
65 mm	$FL = Q^2 + Q$
77 mm	$FL = \frac{Q^2 + Q}{2.5}$
90 mm	$FL = \frac{Q^2 + Q}{5.8}$
100 mm	$FL = \frac{Q^2 + Q}{11}$
115 mm	$FL = \frac{Q^2 + Q}{20}$
130 mm	$FL = \frac{Q^2 + Q}{32}$
140 mm	$FL = \frac{Q^2 + Q}{50}$
160 mm	$FL = \frac{Q^2 + Q}{80}$

where FL = friction loss per 30 m hose
Q = flow/100 L/min

Metric formulas supplied by Office of the Fire Marshal, Toronto, Ontario, Canada.

Required Pump Discharge Pressure

$$PDP = NP + FL$$

where PDP = Pump discharge pressure
NP = Nozzle pressure
FL = Friction loss

PDP with Back Pressure and Pressure "Gain"

$$PDP = NP + FL + 0.05(h)$$

where h = Feet above or below pump
0.05 is a constant

Net Engine Pressure

At Hydrant:

$$NEP = PDP - IP \text{ (in psi)}$$

At Draft:

$$NEP = PDP + FL + IP \text{ (in psi)}$$

Maximum Lift

$$h = 1.13Hg$$

where h = Maximum lift (feet)
Hg = Height of mercury (inches)
1.13 is a constant

TABLE A.1

Theoretical Discharge through Solid Stream Nozzles

Nozzle Pressure		Nozzle Diameter in Inches																			
		1/8	3/16	1/4	3/8	1/2	5/8	3/4	7/8	15/16	1	1 1/8	1 1/4	1 3/8	1 1/2	1 5/8	1 3/4	1 7/8	2	2 1/4	2 1/2
PSI	Feet Head	U.S. Gallons per Minute																			
10	23.09	1.48	3.30	5.90	13.3	23.6	36.8	53.2	72.2	82	94.4	119	148	179	212	248	289	330	378	478	590
15	34.63	1.81	4.02	7.23	16.3	28.7	45.0	65.1	88.4	101	116	146	181	219	260	300	354	400	463	586	722
20	46.18	2.09	4.66	8.35	18.8	33.4	52.0	75.3	102	116	134	169	209	253	300	350	409	468	534	676	834
25	57.72	2.34	5.25	9.34	21.0	37.3	58.3	84.0	114	129	149	189	234	283	336	391	458	521	598	756	934
30	69.27	2.56	5.71	10.2	23.0	40.9	63.7	92.2	125	142	164	207	256	309	368	429	501	572	654	828	1022
35	80.81	2.77	6.21	11.1	24.8	44.2	69.0	99.5	135	153	177	224	277	334	398	464	541	618	708	895	1106
40	92.36	2.95	6.60	11.8	26.5	47.2	73.6	106	144	164	189	239	295	357	425	496	578	661	755	956	1180
45	103.90	3.13	7.03	12.5	28.2	50.1	78.2	113	153	174	200	253	313	379	451	526	613	702	810	1015	1252
50	115.45	3.30	7.37	13.2	29.7	52.8	82.3	119	161	183	211	267	330	399	475	554	646	740	845	1069	1319
55	126.99	3.46	7.77	13.8	31.1	55.3	86.4	125	169	192	221	280	342	418	498	582	678	776	886	1121	1385
60	138.54	3.62	8.08	14.5	32.5	57.8	90.1	130	177	201	231	293	362	438	520	607	708	810	925	1171	1445
65	150.08	3.77	8.45	15.1	33.8	60.2	94.0	136	184	209	241	305	376	455	542	632	737	844	964	1220	1506
70	161.63	3.91	8.73	15.6	35.1	62.5	97.4	141	191	217	250	316	391	473	562	655	765	875	999	1265	1561
75	173.17	4.05	9.08	16.2	36.4	64.7	101	146	198	224	259	327	404	489	582	679	792	906	1037	1310	1619
80	184.72	4.18	9.33	16.7	37.5	66.8	104	150	204	232	267	338	418	505	601	700	818	935	1068	1352	1669
85	196.26	4.31	9.67	17.3	38.8	68.9	108	155	211	238	276	349	431	521	620	722	844	965	1103	1395	1723
90	207.81	4.43	9.89	17.7	39.8	70.8	110	160	217	245	283	358	443	536	637	743	867	992	1133	1434	1770
95	219.35	4.56	10.2	18.2	41.0	72.8	114	164	223	252	292	369	456	551	656	764	892	1019	1168	1476	1824
100	230.90	4.67	10.4	18.7	42.0	74.6	116	168	228	259	299	378	467	565	672	783	914	1043	1194	1512	1866
105	242.44	4.79	10.8	19.2	43.1	76.5	120	172	234	265	306	388	479	579	689	803	937	1070	1226	1550	1916
110	253.98	4.90	10.9	19.6	44.0	78.3	122	177	239	271	313	396	490	593	705	822	959	1095	1253	1586	1957
115	265.53	5.01	11.2	20.0	45.1	80.1	125	180	245	277	320	406	501	606	720	840	980	1120	1282	1621	2005
120	277.07	5.12	11.4	20.4	46.0	81.8	127	184	250	284	327	414	512	619	736	858	1001	1144	1308	1656	2044
125	318.62	5.22	11.7	20.9	47.0	83.5	130	188	256	289	334	423	522	632	751	876	1022	1168	1338	1690	2090
130	300.16	5.32	11.9	21.3	47.9	85.1	133	192	260	295	341	431	532	645	766	893	1042	1191	1362	1724	2127
140	323.26	5.53	12.4	22.1	49.8	88.4	138	199	271	311	354	448	553	668	795	927	1082	1235	1415	1790	2212
150	346.34	5.72	12.9	22.9	51.5	91.5	143	206	280	322	366	463	572	692	824	960	1120	1279	1466	1853	2290

TABLE A.2
Fog Nozzle GPM Discharges at Various Nozzle Pressures

GPM at 100 PSI	Nozzle Pressure in Pounds per Sq. In.														
	30	40	50	60	70	80	90	100	110	120	130	140	150	400	600
	U.S. Gallons per Minute														
9	5	6	6.5	7	7.5	8	8.5	9	9.5	10	10.2	10.5	11	18	22
12	6.5	7.5	8.5	9	10	11	11.5	12	12.5	13	13.5	14	14.5	24	29
14	7.5	9	10	11	11.5	12	13	14	14.5	15	16	16.5	17	28	
15	8	9.5	10.5	11.5	12.5	13	14	15	16	16.5	17	18	18.5	30	
18	10	11	13	14	15	16	17	18	19	20	20.5	21	22	36	
20	11	12.5	14	15	17	18	19	20	21	22	23	24	24.5	40	
21	11.5	13	15	16	18	19	20	21	22	23	24	25	26	42	
23	12.5	14.5	16	18	19	21	22	23	24	25	26	27	28	46	
24	13	15	17	19	20	22	23	24	25	26	27	28	29	48	
27	15	17	19	21	23	24	26	27	28	30	31	32	33	54	
28	15	18	20	22	24	25	27	28	29	31	32	33	34	56	
32	18	20	23	25	27	29	30	32	34	35	36	38	39	64	
34	19	22	24	26	28	30	32	34	36	37	39	40	42		
42	23	27	30	32	35	38	40	42	44	46	48	50	51		
47	26	30	33	36	39	42	45	47	49	51	53	56	57		
50	27	31	35	39	42	45	47	50	52	55	57	59	61		
51	28	32	36	40	43	46	48	51	53	56	58	60	62		
54	29	34	38	42	45	48	51	54	57	59	61	64	66		
55	30	35	39	43	46	49	52	55	58	60	63	65	67		
60	33	38	42	46	50	54	57	60	63	66	68	71	73		
61	33	39	43	47	51	55	58	61	64	67	69	72	75		
65	36	41	46	50	54	58	62	65	68	71	74	77	80		
78	43	49	55	60	65	70	74	78	82	85	89	92	95		
90	49	57	64	70	75	80	85	90	94	99	103	106	110		
92	50	58	65	71	77	82	87	92	96	101	105	109	113		
95	52	60	67	73	79	85	90	95	100	104	108	112	116		
96	53	61	68	74	80	86	91	96	101	105	109	113	117		
100	55	63	71	77	84	89	95	100	105	110	114	118	122		
106	58	67	75	82	88	95	100	100	111	116	121	125	130		
107	59	68	76	83	89	96	101	107	112	117	122	127	131		
108	59	68	77	84	90	97	102	108	113	118	123	128	132		
118	65	75	83	91	99	106	112	118	124	129	135	140	144		
120	66	76	85	93	100	107	114	120	126	131	137	142	147		
125	68	79	88	97	105	112	119	125	131	137	142	148	153		
140	77	89	99	108	117	125	133	140	147	153	160	166	171		
154	84	97	109	119	129	138	146	154	161	169	175	182	189		
240	131	152	170	186	201	215	228	240	252	263	274	284	294		
250	137	158	177	194	209	224	237	250	262	274	285	296	306		
280	153	177	198	217	234	250	266	280	294	307	319	331	343		
400	219	253	283	310	335	358	379	400	420	438	456	473	490		
500	274	316	354	387	418	447	474	500	524	548	570	592	612		
700	383	443	495	542	586	626	664	700	734	767	798	828	857		
1000	548	632	707	775	837	894	949	1000	1049	1095	1140	1183	1225		
1350	739	854	955	1046	1129	1207	1281	1350	1416	1479	1539	1597	1653		
2000	1095	1265	1414	1549	1673	1789	1897	2000	2098	2191	2280	2366	2449		

TABLE A.3

Outlet Pressure Measured by Pitot Gauge

Head Lbs. per Sq. In.	Head Feet	Velocity of Discharge in Ft per Second	Diameter of Orifice in Inches 1/16	1/8	3/16	1/4	3/8	1/2	5/8	3/4	7/8	1	1 1/8	1 1/4	1 1/2	1 3/4	2	2 1/4	2 1/2	2 3/4	3	3 1/4	3 1/2	3 3/4	4	4 1/2	5	6	Head Lbs. per Sq. In.
1	2.31	12.20	.12	.47	1.05	1.87	4.20	7.46	11.7	16.8	22.9	29.9	37.8	46.7	67.2	91.4	119	151	187	226	269	315	366	420	478	604	746	1074	1
2	4.62	17.25	.16	.66	1.48	2.64	5.94	10.5	16.5	23.7	32.3	42.2	53.4	66.0	95.0	129	169	214	264	319	380	446	517	594	676	854	1005	1515	2
3	6.93	21.13	.20	.81	1.82	3.23	7.27	12.9	20.2	29.1	39.6	51.7	65.4	80.8	116	158	207	262	323	391	465	546	633	727	827	1045	1292	1858	3
4	9.24	24.39	.23	.93	2.10	3.73	8.40	14.9	23.3	33.6	45.7	59.7	75.6	93.3	134	183	239	302	373	452	537	631	731	840	955	1210	1492	2148	4
5	11.54	27.26	.26	1.04	2.35	4.17	9.39	16.7	26.1	37.6	51.1	66.7	84.5	104	150	204	267	338	417	505	601	705	817	938	1068	1350	1668	2406	5
6	13.85	29.87	.29	1.14	2.57	4.57	10.3	18.3	28.6	41.1	56.0	73.1	92.5	114	164	224	292	370	457	553	658	772	896	1028	1170	1480	1826	2631	6
7	16.16	32.26	.31	1.23	2.78	4.94	11.1	19.7	30.8	44.4	60.5	79.0	99.9	123	178	242	316	400	494	597	711	834	967	1111	1263	1600	1973	2846	7
8	18.47	34.49	.33	1.32	2.97	5.28	11.9	21.1	33.0	47.5	64.6	84.4	107	132	190	259	338	427	528	638	760	892	1034	1187	1351	1710	2109	3039	8
9	20.78	36.58	.35	1.40	3.15	5.60	12.6	22.4	35.0	50.4	68.6	89.5	113	140	201	274	358	453	560	677	806	946	1097	1259	1433	1815	2237	3222	9
10	23.09	38.56	.37	1.48	3.30	5.90	13.3	23.6	36.8	53.2	72.2	94.4	119	148	212	289	378	478	590	714	850	997	1156	1327	1510	1910	2358	3394	10
11	25.40	40.45	.39	1.55	3.45	6.19	13.9	24.7	38.6	55.8	75.7	99.0	125	155	223	303	396	501	610	749	891	1046	1213	1392	1584	2010	2474	3565	11
12	27.71	42.24	.40	1.62	3.60	6.47	14.5	25.7	40.3	58.3	79.1	103	131	162	233	317	414	524	646	782	931	1092	1267	1454	1655	2100	2584	3716	12
13	30.02	43.97	.42	1.68	3.74	6.73	15.1	26.7	41.9	60.7	82.3	108	136	168	242	330	431	545	673	814	969	1137	1318	1515	1722	2180	2689	3877	13
14	32.33	45.63	.43	1.75	3.88	6.98	15.7	27.7	43.5	63.0	85.4	112	141	175	251	342	447	566	698	845	1005	1180	1368	1572	1787	2260	2791	4016	14
15	34.63	47.22	.45	1.81	4.02	7.23	16.3	28.7	45.0	65.1	88.4	116	146	181	260	354	463	586	722	874	1040	1221	1416	1626	1849	2340	2888	4156	15
16	36.94	48.78	.46	1.87	4.16	7.47	16.8	29.7	46.5	67.2	91.3	120	151	187	269	366	478	605	746	903	1075	1261	1463	1679	1910	2420	2983	4296	16
17	39.25	50.28	.48	1.92	4.29	7.69	17.3	30.7	47.9	69.3	94.1	123	156	192	277	377	493	623	769	931	1108	1300	1508	1731	1969	2500	3075	4424	17
18	41.56	51.73	.49	1.98	4.42	7.92	17.8	31.6	49.3	71.3	96.8	127	160	198	285	388	507	642	791	958	1140	1338	1551	1781	2026	2570	3164	4553	18
19	43.87	53.15	.51	2.03	4.54	8.13	18.3	32.5	50.7	73.3	99.5	130	165	203	293	399	521	659	813	984	1171	1374	1594	1830	2082	2640	3251	4682	19
20	46.18	54.54	.52	2.09	4.66	8.35	18.8	33.4	52.0	75.3	102	134	169	209	300	409	534	676	834	1010	1201	1410	1635	1877	2136	2710	3335	4800	20
22	50.80	57.19	.55	2.19	4.90	8.75	19.7	35.0	54.6	78.9	107	140	177	219	315	429	560	709	875	1059	1260	1479	1715	1969	2240	2840	3498	5037	22
24	55.41	59.74	.57	2.29	5.13	9.14	20.6	36.4	57.0	82.4	112	146	185	229	329	448	585	741	914	1106	1316	1545	1791	2056	2340	2970	3654	5262	24
26	60.03	62.18	.59	2.38	5.34	9.52	21.4	38.0	59.4	85.8	116	152	193	238	343	466	609	771	951	1151	1370	1608	1864	2140	2435	3090	3803	5477	26
28	64.65	64.52	.62	2.47	5.53	9.88	22.2	39.5	61.6	89.0	121	158	200	247	356	484	632	800	987	1194	1422	1668	1935	2221	2527	3210	3946	5681	28
30	69.27	66.79	.64	2.56	5.71	10.2	23.0	40.9	63.7	92.2	125	164	207	256	368	501	654	828	1022	1236	1472	1727	2003	2299	2616	3320	4087	5885	30
32	73.89	68.98	.66	2.64	5.90	10.6	23.8	42.3	65.8	95.2	129	169	214	264	380	517	676	856	1055	1277	1520	1784	2069	2375	2702	3430	4221	6078	32
34	78.50	71.10	.68	2.72	6.07	10.9	24.5	43.6	67.8	98.2	133	174	220	272	392	533	697	882	1088	1316	1566	1838	2132	2448	2785	3540	4348	6261	34
36	83.12	73.16	.70	2.80	6.25	11.2	25.2	44.8	69.8	101	137	179	226	280	403	548	717	908	1119	1354	1612	1892	2194	2519	2866	3640	4475	6443	36
38	87.74	75.17	.72	2.88	6.42	11.5	25.9	46.0	71.8	104	141	184	233	288	414	563	736	932	1150	1392	1656	1944	2254	2588	2944	3740	4594	6615	38
40	92.36	77.11	.74	2.95	6.60	11.8	26.5	47.2	73.6	106	144	189	239	295	425	578	755	956	1180	1428	1699	1994	2313	2655	3021	3840	4713	6787	40
42	96.98	79.03	.75	3.03	6.76	12.1	27.2	48.4	75.4	109	148	193	245	303	435	592	774	980	1209	1463	1741	2043	2370	2721	3095	3935	4833	6959	42
44	101.59	80.88	.77	3.10	6.92	12.4	27.8	49.6	77.2	112	151	198	251	310	445	606	792	1003	1237	1497	1782	2091	2426	2785	3168	4030	4945	7120	44
46	106.21	82.70	.79	3.17	7.07	12.7	28.5	50.8	79.0	114	155	203	256	317	455	620	810	1025	1265	1531	1822	2138	2480	2847	3239	4120	5056	7281	46
48	110.83	84.48	.81	3.24	7.22	12.9	29.1	51.8	80.7	117	158	207	262	324	465	633	828	1047	1293	1564	1861	2184	2533	2908	3309	4205	5168	7442	48
50	115.45	86.22	.82	3.30	7.37	13.2	29.7	52.8	82.3	119	161	211	267	330	475	646	845	1069	1319	1596	1900	2229	2586	2968	3377	4290	5273	7592	50

52	120.07	87.93	.84	3.37	7.52	13.5	30.3	53.8	83.9	121	165	215	272	337	485	659	861	1091	1345	1628	1937	2274	2637	3027	3444	4375	5377	7743	52
54	124.68	89.61	.85	3.43	7.66	13.7	30.9	54.9	85.5	124	168	219	277	343	494	672	878	1111	1371	1659	1974	2317	2687	3085	3510	4460	5482	7893	54
56	129.30	91.20	.87	3.50	7.80	14.0	31.4	55.9	87.1	126	171	223	283	350	503	684	894	1132	1396	1689	2010	2359	2736	3141	3574	4540	5579	8033	56
58	133.92	92.87	.89	3.56	7.94	14.2	32.0	56.9	88.6	128	174	227	288	356	512	696	909	1152	1421	1719	2046	2401	2785	3197	3637	4620	5683	8183	58
60	138.54	94.45	.90	3.62	8.08	14.5	32.5	57.8	90.1	130	177	231	293	362	520	708	925	1171	1445	1749	2081	2442	2832	3252	3700	4700	5780	8323	60
62	143.15	96.01	.92	3.68	8.21	14.7	33.0	58.8	91.6	133	180	235	297	368	529	720	941	1191	1470	1777	2115	2483	2879	3305	3761	4775	5869	8452	62
64	147.77	97.55	.93	3.74	8.34	14.9	33.6	59.7	93.1	135	183	239	302	374	538	731	956	1210	1493	1806	2149	2522	2925	3358	3821	4850	5966	8591	64
66	152.39	99.07	.95	3.79	8.47	15.2	34.1	60.6	94.6	137	186	243	307	379	546	742	971	1228	1516	1834	2183	2561	2971	3410	3880	4925	6056	8720	66
68	157.01	100.55	.96	3.85	8.59	15.4	34.6	61.5	96.0	139	188	246	311	385	554	754	985	1247	1539	1862	2215	2600	3015	3462	3938	5000	6153	8860	68
70	161.63	102.03	.97	3.91	8.73	15.6	35.1	62.5	97.4	141	191	250	316	391	562	765	999	1265	1561	1889	2248	2638	3059	3512	3996	5075	6242	8989	70
72	166.24	103.47	.99	3.96	8.85	15.8	35.6	63.3	98.7	143	194	253	320	396	570	776	1014	1283	1583	1916	2280	2675	3103	3562	4053	5140	6332	9117	72
74	170.86	104.90	1.00	4.02	8.97	16.0	36.1	64.2	100	145	196	257	325	402	578	786	1028	1301	1605	1942	2311	2712	3146	3611	4109	5200	6414	9236	74
76	175.48	106.30	1.01	4.07	9.09	16.3	36.6	65.1	101	147	199	260	329	407	586	797	1041	1318	1627	1968	2342	2749	3188	3660	4164	5265	6503	9364	76
78	180.10	107.69	1.03	4.12	9.21	16.5	37.1	65.9	103	149	202	264	333	412	593	807	1055	1335	1648	1994	2373	2785	3230	3708	4218	5340	6585	9483	78
80	184.72	109.08	1.04	4.18	9.33	16.7	37.5	66.8	104	150	204	267	338	418	601	818	1068	1352	1669	2019	2403	2820	3271	3755	4272	5405	6667	9601	80
82	189.33	110.42	1.05	4.23	9.45	16.9	38.0	67.6	105	152	207	270	342	423	608	828	1082	1369	1689	2044	2433	2855	3311	3801	4325	5476	6757	9730	82
84	193.95	111.76	1.07	4.28	9.56	17.1	38.4	68.4	107	154	209	274	346	428	616	838	1095	1386	1710	2069	2462	2890	3351	3847	4377	5535	6839	9848	84
86	198.57	113.08	1.08	4.33	9.68	17.3	38.9	69.2	108	156	212	277	350	433	624	848	1107	1402	1730	2094	2491	2924	3391	3893	4429	5600	6914	9955	86
88	203.19	114.39	1.09	4.38	9.79	17.5	39.3	70.0	109	158	214	280	354	438	630	858	1120	1419	1750	2118	2520	2958	3430	3938	4480	5665	6996	10073	88
90	207.81	115.68	1.10	4.43	9.89	17.7	39.8	70.8	110	160	217	283	358	443	637	867	1133	1434	1770	2142	2549	2991	3469	3983	4531	5730	7078	10191	90
92	212.42	116.96	1.12	4.48	10.0	17.9	40.2	71.6	112	161	219	286	362	448	644	877	1146	1450	1789	2165	2577	3024	3507	4027	4581	5795	7152	10299	92
94	217.04	118.23	1.13	4.53	10.1	18.1	40.7	72.4	113	163	221	289	366	453	651	886	1158	1466	1809	2189	2605	3057	3545	4070	4631	5865	7234	10417	94
96	221.66	119.48	1.14	4.57	10.2	18.3	41.1	73.2	114	165	224	293	370	457	658	895	1170	1481	1828	2212	2632	3089	3583	4113	4680	5925	7309	10524	96
98	226.28	120.71	1.15	4.62	10.3	18.5	41.6	73.9	115	167	226	296	374	462	665	904	1182	1497	1847	2235	2660	3121	3620	4156	4728	5985	7383	10632	98
100	230.90	121.94	1.16	4.67	10.4	18.7	42.0	74.6	116	168	228	299	378	467	672	914	1194	1512	1866	2258	2687	3153	3657	4198	4776	6045	7458	10739	100
102	235.51	123.15	1.18	4.72	10.5	18.8	42.4	75.4	118	170	231	302	381	472	679	923	1206	1527	1884	2280	2713	3184	3693	4240	4824	6100	7533	10846	102
104	240.13	124.35	1.19	4.76	10.6	19.0	42.8	76.1	119	172	233	304	385	476	685	932	1218	1542	1903	2302	2740	3215	3729	4281	4871	6150	7607	10954	104
106	244.75	125.55	1.20	4.81	10.7	19.2	43.4	76.9	120	173	235	307	389	481	692	941	1230	1556	1921	2324	2766	3246	3765	4322	4917	6200	7682	11061	106
108	249.37	126.73	1.21	4.85	10.8	19.4	43.6	77.6	121	175	237	310	392	485	698	950	1241	1571	1939	2346	2792	3277	3800	4363	4963	6260	7749	11158	108
110	253.98	127.89	1.22	4.90	10.9	19.6	44.0	78.3	122	177	239	313	396	490	705	959	1253	1586	1957	2368	2818	3307	3835	4403	5009	6320	7823	11265	110
112	258.60	129.05	1.23	4.94	11.0	19.7	44.4	79.0	123	178	242	316	400	494	711	967	1264	1600	1974	2389	2843	3337	3870	4443	5054	6380	7891	11362	112
114	263.22	130.20	1.24	4.99	11.1	19.9	44.8	79.7	124	180	244	319	403	499	717	976	1275	1614	1992	2410	2869	3367	3904	4482	4099	6440	7965	11469	114
116	267.84	131.33	1.25	5.03	11.2	20.1	45.2	80.4	125	131	246	322	407	503	724	984	1286	1628	2009	2431	2894	3396	3938	4521	5144	6500	8032	11566	116
118	272.46	132.46	1.26	5.07	11.3	20.3	45.6	81.1	126	133	248	324	410	507	730	993	1297	1642	2027	2452	2918	3425	3972	4560	5188	6560	8099	11663	118
120	277.07	133.57	1.27	5.12	11.4	20.4	46.0	81.8	127	184	250	327	414	512	736	1001	1308	1656	2044	2473	2943	3454	4006	4599	5232	6620	8167	11759	120
122	281.69	134.69	1.29	5.16	11.5	20.6	46.4	82.5	129	186	252	330	417	516	742	1010	1319	1670	2061	2494	2967	3483	4039	4637	5275	6680	8241	11867	122
124	286.31	135.79	1.30	5.20	11.6	20.8	46.7	83.1	130	187	254	333	421	520	748	1018	1330	1684	2077	2514	2992	3511	4072	4675	5318	6740	8308	11963	124
126	290.93	136.88	1.31	5.24	11.7	20.9	47.1	83.8	131	189	256	335	424	524	754	1026	1341	1697	2094	2534	3016	3539	4105	4712	5361	6800	8375	12060	126
128	295.55	137.96	1.32	5.28	11.8	21.1	47.5	84.5	132	190	258	338	427	528	760	1034	1351	1711	2111	2554	3040	3567	4137	4749	5403	6850	8435	12146	128
130	300.16	139.03	1.33	5.32	11.9	21.3	47.9	85.1	133	192	260	341	431	532	766	1042	1362	1724	2127	2574	3063	3595	4169	4786	5445	6900	8502	12242	130
132	304.78	140.10	1.34	5.36	12.0	21.4	48.2	85.7	134	193	263	343	434	536	772	1050	1372	1736	2144	2594	3087	3623	4201	4823	5487	6950	8569	12339	132
134	309.40	141.16	1.35	5.40	12.1	21.6	48.6	86.4	135	194	265	346	437	540	777	1058	1382	1749	2160	2613	3110	3650	4233	4860	5529	7000	8636	12436	134
136	314.02	142.21	1.36	5.44	12.2	21.8	49.0	87.0	136	196	267	348	441	544	783	1066	1392	1762	2176	2633	3133	3677	4265	4896	5570	7050	8695	12522	136

TABLE A.4
Friction Loss in Rubber-Lined Fire Hose
(Using $2Q^2 + Q$)

	Diameter (inches) Single Lines											Siamesed Lines Equal Length		
GPM	¾	1	1½	1¾	2	2½	3	3½	4	5	6	Two 2½	Three 2½	Two 3
10	24.08	9.18	0.95	0.42	0.21									
20	61.92	16.38	2.43	1.07	0.53									
30	113.52	30.03	4.46	1.96	0.97									
40		47.32	7.02	3.09	1.53									
50		68.25	10.13	4.46	2.21									
60			13.77	6.07	2.99									
70			17.96	7.91	3.91									
80			22.68	9.99	4.94									
90			27.95	12.32	6.09									
100			40.50	17.85	8.82	3.00	1.16							
110			47.52	20.94	10.35	3.52	1.36					0.99		
120			55.08	24.28	11.99	4.08	1.57					1.14		
130				27.85	13.76	4.68	1.80					1.31		
140				31.65	15.64	5.32	2.05					1.49		
150				35.70	17.64	6.00	2.31					1.68		
160				39.98	19.76	6.72	2.59					1.88		
170				44.51	21.99	7.48	2.88	1.29				2.09	0.96	0.80
180				49.27	24.34	8.28	3.19	1.42				2.31	1.07	0.89
190				54.26	26.81	9.12	3.51	1.56				2.55	1.18	0.98
200					29.40	10.00	3.85	1.72	0.90			2.80	1.29	1.07
210					32.10	10.92	4.20	1.87	0.98			3.05	1.41	1.17
220					34.93	11.88	4.57	2.04	1.06			3.32	1.53	1.27
230					37.87	12.88	4.95	2.21	1.15			3.60	1.66	1.38
240					40.92	13.92	5.35	2.39	1.25			3.89	1.79	1.49
250					44.10	15.00	5.77	2.58	1.35			4.20	1.93	1.61
260					47.39	16.12	6.20	2.77	1.45			4.51	2.07	1.72
270					50.80	17.28	6.65	2.97	1.55			4.83	2.22	1.85
280						18.48	7.11	3.17	1.66			5.17	2.38	1.98
290						19.72	7.59	3.39	1.77			5.52	2.54	2.11
300						21.00	8.08	3.61	1.89			5.88	2.70	2.25
310						22.32	8.59	3.83	2.00			6.24	2.87	2.39

	Single Lines						Siamesed Lines Equal Length		
320	23.68	9.11	4.07	2.13			6.63	3.05	2.53
330	25.08	9.65	4.31	2.25			7.02	3.23	2.68
340	26.52	10.21	4.56	2.38			7.42	3.42	2.84
350	28.00	10.78	4.81	2.52	0.87		7.84	3.61	2.99
360	29.52	11.36	5.07	2.65	0.91		8.26	3.80	3.15
370	31.08	11.96	5.34	2.79	0.96		8.70	4.00	3.32
380	32.68	12.58	5.62	2.94	1.01		9.15	4.21	3.49
390	34.32	13.21	5.90	3.08	1.06		9.60	4.38	3.67
400	36.00	13.86	6.19	3.24	1.16		10.08	4.64	3.85
425	40.38	15.54	6.94	3.63	1.25		11.30	5.20	4.32
450	45.00	17.32	7.74	4.05	1.39		12.60	5.80	4.81
475	49.88	19.20	8.57	4.48	1.54		13.96	6.43	5.33
500	55.00	21.17	9.46	4.95	1.70		15.40	7.09	5.88
525		23.24	10.38	5.43	1.87		16.90	7.78	6.46
550		25.41	11.35	5.94	2.04		18.48	8.50	7.06
575		27.67	12.36	6.46	2.22	0.86	20.12	9.27	7.69
600		30.03	13.41	7.02	2.41	0.93	21.84	10.06	8.34
625		32.48	14.51	7.59	2.61	1.01	23.62	10.88	9.02
650		35.03	15.65	8.19	2.82	1.09	25.48	11.73	9.73
675		37.68	16.83	8.80	3.03	1.17	27.40	12.62	10.47
700		40.42	18.06	9.45	3.25	1.26	29.40	13.54	11.23
725		43.26	19.32	10.11	3.48	1.34	31.46	14.49	12.02
750		46.10	20.59	10.77	3.71	1.43	33.53	15.44	12.81
775		49.23	21.99	11.50	3.96	1.53	35.80	16.49	13.68
800		52.36	23.39	12.24	4.21	1.63	38.08	17.54	14.55
825			24.83	12.99	4.47	1.73	40.42	18.62	15.44
850			26.31	13.77	4.74	1.83	42.84	19.73	16.37
875			27.84	14.56	5.01	1.94	45.32	20.88	17.32
900			29.41	15.39	5.30	2.05	47.88	22.05	18.29
925			31.02	16.23	5.60	2.16	50.50	23.26	19.30
950			32.68	17.10	5.89	2.28		24.51	20.33
975			34.37	17.98	6.19	2.39		25.78	21.38
1,000			36.12	18.90	6.51	2.52		27.09	22.47
1,050			39.73	20.79	7.16	2.77		29.79	24.71
1,100			43.51	22.77	7.84	3.03		32.63	27.07
1,150			47.47	24.84	8.55	3.31		35.60	29.53
1,200			51.60	27.00	9.30	3.60		38.70	32.10
1,250				29.25	10.07	3.90		41.92	34.77
1,500				41.85	14.41	5.58		59.98	49.75
2,000				73.80	25.42	9.84			87.74

TABLE A.5
Relative Discharge Capacities of Pipes Flowing Full

Diam. in Inches	Diameter in Inches								
	4	6	8	12	16	20	24	30	36
36				15.58	7.59	4.34	2.74	1.58	1
30			27.23	9.85	4.8	2.74	1.74	1	
24		32.00	15.58	5.65	2.75	1.57	1		
20	55.96	20.29	9.88	3.58	1.74	1			
16	32.01	11.6	5.65	2.05	1				
12	15.60	5.65	2.75	1					
8	5.66	2.05	1						
6	2.75	1							

TABLE A.6
Minimum Pump Pressure, Lines and Nozzles for Good Flow

Normal Output of Pump		Minimum Pump Pressure for 80% of Nominal Output	Number of Lines	Number and Sizes of Nozzles (mm)
at 700 kPa	At 1,000 kPa (U.S. pumps)			
95 L/s	50	550	5	Five 25
		700	3	Two 25 One 22
80 L/s	45	550	5	Two 25 Three 22
		700	3	Three 22
55-60 L/s	30-33	550	3	Three 25
		700	3	Three 20
35-40 L/s	25	550	2	Two 25
		700	2	One 22 One 20
25-30 L/s		550	2	One 22 One 20
		700	1	One 25
18-20 L/s	12-15	550	2	Two 20
		700	1	One 22
15 L/s		500	2	Two 16
5-8 L/s		700	1	One 16

Note: When the nominal output of the pump falls between any two figures in the table, the number and/or size of the nozzles should be adjusted accordingly.

Source: New Zealand Fire Service, *Drill Book*, part I, 1979 edition.

LIST OF TABLES

INDEX

SPACE FOR CALCULATIONS AND NOTES

SPACE FOR CALCULATIONS AND NOTES

IFSTA MATERIALS

101 FORCIBLE ENTRY, ROPE AND PORTABLE EXTINGUISHER PRACTICES
Types of forcible entry tools and general building construction; use of tools in opening doors, windows, roofs, floors, walls, partitions and ceilings; types, uses, and care of ropes, knots and portable fire extinguishers.

102 FIRE SERVICE GROUND LADDER PRACTICES
Various terms applied to ladders; types, construction, maintenance, and testing of fire service ground ladders; detailed information on handling ground ladders and special tasks related to them.

103 FIRE HOSE PRACTICES
Construction, care, and testing of hose and various fire hose accessories; preparation and manipulation of hose for rolls, folds, connections, carries, drags, and special operations; loads and layouts for fire hose.

104 SALVAGE AND OVERHAUL PRACTICES
Planning and preparing for salvage operations, care and preparation of equipment, methods of spreading and folding salvage covers, most effective way to handle water runoff, value of proper overhaul and equipment needed, determining cause of fire, and recognizing and preserving arson evidence.

105 FIRE STREAM PRACTICES
Characteristics, requirements and principles of fire streams; developing, computing, and applying various types of streams to operational situations; formulas for application of hydraulics; actions and reactions created by applying streams under different circumstances.

106 FIRE APPARATUS PRACTICES
Various types of fire apparatus classified by functions; driving and operating apparatus including pumpers, aerial ladders, and elevating platforms; maintenance and testing of apparatus.

107 FIRE VENTILATION PRACTICES
Objectives and advantages of ventilation; requirements for burning, flammable liquid characteristics and products of combustion; phases of burning, backdrafts, and the transmission of heat; construction features to be considered; the ventilation process including evaluating and size-up is discussed in length.

108 FIRE SERVICE RESCUE AND PROTECTIVE BREATHING PRACTICES
Respiratory tract and respiratory hazards; protective breathing equipment and how to use, care, test, and inspect; other protection equipment, including clothing, ambulances, and rescue equipment; rescue methods, techniques, procedures and situations.

109 FIRE SERVICE FIRST AID PRACTICES
Completely revised. Brief explanations of the nervous, skeletal, muscular, abdominal, digestive, and genitourinary systems; injuries and treatment relating to each system; bleeding control and bandaging; artificial respiration, external cardiac compression and cardiopulmonary resuscitation; shock, poisoning, and emergencies caused by heat and cold; fractures, sprains, and dislocations; emergency childbirth; short-distance transfer of patients and ambulances; conducting a primary and secondary survey.

110 FIRE PREVENTION AND INSPECTION PRACTICES
Fire prevention bureau and inspecting agencies; fire hazards and causes; prevention and inspection techniques; building construction, occupancy and fire load; special-purpose inspections; inspection forms and checklists along with reference sources; maps and symbols; records and reports.

200 ESSENTIALS OF FIRE FIGHTING
This new manual was prepared to meet the objectives set forth in levels I and II of NFPA, *Fire Fighter Professional Qualifications, 1974.* Included in the manual are the basics of: hose, ladders, fire streams, forcible entry, rescue, salvage and overhaul, ventilation, fire behavior and science, ropes and knots, extinguishers, protective breathing, fire prevention, fire cause identification, ground cover, communications, water supplies and sprinkler systems.

200 SI SELF-INSTRUCTION FOR IFSTA 200
Self-instruction book for IFSTA 200 *Essentials of Fire Fighting.* Covers the most important points of IFSTA 200, including NFPA 1001 Firefighter I and II requirements. Perfect supplement for formal training.

201 FIRE SERVICE PRACTICES FOR VOLUNTEER FIRE DEPARTMENTS
A general, cursory introduction to material covered in detail in manuals 101 - 110, with the exception of first aid.

202 FIRE SERVICE ORIENTATION & INDOCTRINATION
History, traditions, and organization of the fire service; operation of the fire department and responsibilities and duties of firefighters; fire department companies and their functions; glossary of fire service terms.

203 FIRE SERVICE TRAINING PROGRAMS
Qualifications and duties of the training officer in planning, organizing, directing, and controlling the training program; examples of training centers, facilities, and programs; training curricula — a master outline of hundreds of address sources.

204 PHOTOGRAPHY FOR THE FIRE SERVICE
Camera components and operations, films — their advantages and differences, fundamental principles in taking a good picture, processing the negative and print, controlling light for optimum results, accountability and storage of photographic materials; fire scene photography, using photographs as training aids, enhancing fire prevention and public relations through photography, how the investigator can use photographs, glossary of photographic terms.

205 WATER SUPPLIES FOR FIRE PROTECTION
Importance, basic components, adequacy, reliability, and carrying capacity of water systems; specifications, installation, maintenance and distribution of fire hydrants; flow requirements, flow tests and control valves; sprinkler and standpipe systems.

206 AIRCRAFT FIRE PROTECTION AND RESCUE PROCEDURES
Aircraft types, engines, and systems, conventional and specialized fire fighting apparatus, tools, clothing, extinguishing agents, dangerous materials, communications, pre-fire planning, and airfield operations.

207 GROUND COVER FIRE FIGHTING PRACTICES
Ground cover fire apparatus, equipment, extinguishing agents, and fireground safety; organization and planning for ground cover fire; authority, jurisdiction, and mutual aid, techniques and procedures used for combating ground cover fire.

208 RECORDS AND REPORTS FOR THE FIRE SERVICE
Procedures for compiling information for records, methods of writing and presenting reports, various sample forms for all activities of the fire service.

209 FIREFIGHTER SAFETY
Basic concepts and philosophy of accident prevention; essentials of a safety program and training for safety; station house facility safety; hazards enroute and at the emergency scene; personal protective equipment; special hazards, including chemicals, electricity, and radioactive materials; inspection safety; health considerations.

210 PRIVATE FIRE PROTECTION & DETECTION
Automatic sprinkler systems, special extinguishing systems, standpipes, detection and alarm systems. Includes how to test sprinkler systems for the firefighter to meet NFPA 1001.

301 THE FIRE DEPARTMENT OFFICER
The officer's functions analyzed; decisions, planning, activating, problem solving, delegating authority and supervision; fire fighting activities of the fire officer.

302 FIRE DEPARTMENT — FACILITIES, PLANNING AND PROCEDURES
Fire protection facilities and water supplies; pre-fire planning and fire fighting procedures; dispatching companies, size-up, attack, confinement, extinguishment and overhaul.

303 FIRE SERVICE INSTRUCTOR TRAINING
Characteristics of good instructor; determining training requirements and what to teach; types, principles, and procedures of teaching and learning; training aids and devices; conference leadership.

304 FIRE PROBLEMS IN HIGH-RISE BUILDINGS
Locating, confining, and extinguishing fires; heat, smoke, fire gases, and life hazards; exposures, water supplies and communications; pre-fire planning, ventilation, salvage and overhaul; smokeproof stairways and problems of building design and maintenance; tactical checklist.

401 FUNDAMENTAL PRINCIPLES OF MATHEMATICS
Mathematics oriented to the fire service; fractions, decimals, measurements and weights, percentages and graphs, ratio, proportion, powers, and roots, algebraic and geometric fundamentals and problems.

402 FUNDAMENTAL PRINCIPLES OF SCIENCE
A basic investigation of matter, motion, force, machines, liquids, gases; principles of chemistry, combustion and heat, magnetism, electricity, atomic energy and radiation.

403 LEADERSHIP IN THE FIRE SERVICE
A series of lectures by Robert F. Hamm covering management, supervision, personnel administration, problem solving, and analysis.

500 FIREFIGHTER STUDY GUIDE
Correlates IFSTA publications with NFPA Standard 1001 to guide personnel through the requirements for national firefighter certification at levels I, II, III. Also provides a personal logbook for recording test dates, scores and instructor initials.

502 FIRE SERVICE REFERENCE GUIDE
Each IFSTA manual, except 201 and 403, is individually subject indexed and cross-indexed by general subject; articles from *Fire Command, Fire Engineering, Fire Chief, Fire Journal, W.N.Y.F.* and *Fire News* indexed by subject 1970-76; index of FDIC Proceedings for six years. Separate 1978 *Addendum* indexes an additional year of reference material, plus subject index of IFSTA 101 and 206.

606 PUBLIC FIRE EDUCATION
A valuable contribution to your community's fire safety. Includes public fire education planning, target audiences, seasonal fire problems, smoke detectors, working with the media, burn injuries, and resource exchange.

FIRE PROTECTION ADMINISTRATION
A reprint of the Illinois Department of Commerce and Community Affairs publication. A manual for trustees, municipal officials, and fire chiefs of fire districts and small communities. Subjects covered include officials' duties and responsibilities, organization and management, personnel management and training, budgeting and finance, annexation and disconnection.

INSTRUCTOR GUIDE SETS

Available for manuals 101 - 110, 200, 206, and for the slide program *Fire Department Support of Automatic Sprinkler Systems.* Basic lesson plan, tips for instructor, references. 200 has NFPA Standard 1001 references and pertinent proficiency tests.

SLIDES

Ladder ***Salvage** **Sprinkler**

2-inch by 2-inch slides that can be used in any 35 mm slide projector; supplements to respective manuals and sprinkler guide sets.

*The complete package consists of the slides, instructor's manual, and instructor's guide sets.

Smoke Detectors Can Save Your Life
How To Inspect A Restaurant
Matches Aren't For Children
Public Relations for the Fire Service
Danger! Fire Fighters at Work, Safety 1 (NFPA)
The Fire Department Safety Program, Safety 2 (NFPA)

TRANSPARENCIES

Multi-colored overhead transparencies to augment IFSTA 200, *Essentials of Fire Fighting,* are now available. Since costs and availability vary with different chapters, contact IFSTA Headquarters for details.

MANUAL BINDERS
Polyflex, expandable-pole binders hold 3 - 10 manuals.

GUIDE SHEET BINDERS
Free with purchase of basic 12 guide sets (101-110, 206, and Sprinklers).

WATER FLOW TEST SUMMARY SHEETS
50 summary sheets and instructions on how to use; logarithmic scale to simplify the process of determining the available water in an area.

PERSONNEL RECORD FOLDERS
Personnel record folders should be used by the training officer for each member of the department. Such data as IFSTA training, technical training (seminars), and college work can be recorded in this file, along with other valuable information. Letter size or legal size.